Robotics

Engineering the Future: Innovations and Developments in Robotics

Sarah Reynolds

Table of Contents

INTRODUCTION

In the intricate tapestry of technological evolution, Robotics emerges as a pivotal thread, weaving through the fabric of innovation and shaping the very contours of our future. As we stand at the nexus of man and machine, the world of Robotics unfolds, offering a symphony of possibilities and challenges. This section, titled "Robotics: Engineering the Future," embarks on a compelling journey through the labyrinth of advancements, exploring the profound impact of Robotics on society, industries, and our collective trajectory into the unknown.

Engineering the Future: Innovations and Developments in Robotics is more than a mere exploration of nuts, bolts, and circuits. It is a chronicle of the extraordinary, charting the evolution of robotics from its humble beginnings to the forefront of cutting-edge technologies. As we delve into the fundamentals, the pulsating heartbeat of robotics—comprising mechanics, electronics, and computer science—reveals itself. The various types of robots, from the stalwart industrial machines to the autonomous explorers, are dissected to unveil their unique roles in reshaping our daily lives.

This odyssey through the world of robotics wouldn't be complete without acknowledging the transformative milestones that have paved the way. From the first clunky industrial arm to the sleek, intelligent humanoid, the journey is marked by leaps in artificial intelligence, machine learning, and sensor technologies. The ripple effects of these innovations touch every corner of our existence, influencing industries, healthcare, and even venturing into the vastness of space.

But with great power comes great responsibility. The book navigates through the challenges and ethical considerations that accompany the rise of robotics, addressing concerns about job displacement, safety, and the ethical use of these technologies. As we stand at this crossroads, it becomes imperative to examine not only the possibilities but also the potential pitfalls.

In the chapters that follow, we will scrutinize the latest developments in robotics—dipping our toes into the realms of artificial intelligence, advanced materials, and emerging technologies. Real-world applications, from the heart of manufacturing to the precision of medical robotics, will illuminate the transformative impact of these machines on our lives.

Embark with us on this exploration of the future, where swarm robotics, soft robotics, and human-robot collaboration redefine the boundaries of what is possible. This section is more than a collection of words; it is an invitation to witness the unfolding saga of Robotics: Engineering the Future. Together, let's navigate the uncharted territories of innovation and step boldly into the future that robotics is diligently engineering for us all.

CHAPTER I

Foundations of Robotic Engineering

Core Principles and Concepts

The field of Robotic Engineering stands as a testament to the boundless ingenuity of human minds and the convergence of various scientific disciplines. At its core lie fundamental principles that propel the creation, evolution, and application of robotic systems. These principles form the bedrock upon which the entire discipline rests, guiding engineers and researchers through the intricate journey of designing machines that seamlessly interact with the world around them.

Mechanics, electronics, and computer science stand as the three pillars supporting the foundation of robotic engineering. In the realm of mechanics, the essence of robotic movement takes shape. This encompasses the study of rigid bodies and the dynamics governing their motion. Engineers delve into kinematics, understanding the geometry of motion, and dynamics, unraveling the forces and torques influencing movement. These principles dictate how robotic arms articulate, how wheels turn, and how limbs interact with their environment. Mastery of mechanics is essential for engineers to craft robots that not only move with precision but also respond adeptly to the complexities of real-world scenarios.

Electronics emerge as the vital nervous system of robotic entities. The realm of electronics encompasses the study and application of electrical circuits, sensors, and actuators. Electronic components imbue robots with the ability to perceive their surroundings, process information, and execute actions. Sensors, ranging from vision systems to tactile detectors, serve as the eyes and ears of robots, enabling them to collect data from the environment. Actuators, such as motors and servos, translate electronic signals into physical motion, giving life to the mechanical structure. The synergy between mechanics and electronics transforms robots from inert machines into dynamic entities capable of sensing, processing, and responding in real-time.

Computer science, the third cornerstone, provides the cognitive capacity to robotic systems. Algorithms and software engineering become the programming languages of these mechanical and electronic entities. Artificial intelligence (AI) and machine learning (ML) algorithms empower robots with the ability to learn from experience, make decisions, and adapt to changing circumstances. Computer vision algorithms allow robots to interpret visual data, while path-planning algorithms enable them to navigate complex environments. The infusion of computer science into robotic engineering opens avenues for autonomous operation, enabling robots to perform tasks with minimal human intervention.

As we navigate the intricate landscape of robotic engineering, understanding the types of robots becomes paramount. Industrial robots, with their precision and strength, dominate manufacturing lines, automating processes and enhancing efficiency. Service robots, designed for interaction with humans, find applications in healthcare, hospitality, and domestic settings. Autonomous robots, equipped with sensors and AI, navigate and operate independently, from drones surveying landscapes to rovers exploring distant planets. Each type serves a unique purpose, reflecting the versatility and adaptability of robotic engineering to diverse contexts.

The evolution of robotics unfolds against the backdrop of significant milestones that have shaped the trajectory of the field. The inception of the first industrial robot in the 20th century marked a turning point, heralding an era where machines could perform repetitive tasks with unparalleled precision. Advances in AI and machine learning in recent decades have ushered in a new era, where robots not only follow programmed instructions but also learn and adapt, blurring the lines between machine and human intelligence. The emergence of humanoid robots represents the pinnacle of robotic mimicry, as engineers strive to replicate human form and function.

While the impact of robotics on industries is undeniable, its influence extends far beyond factory floors. Medical robotics has revolutionized healthcare, with robotic surgical systems offering unprecedented precision and minimally invasive procedures. Rehabilitation robotics aids in the recovery of patients, providing personalized therapy and support. Robotic exploration ventures into the depths of oceans and the vastness of space, extending the reach of human presence and knowledge. The evolution of robotics intertwines with the very fabric of modern society, reshaping how we work, live, and explore the unknown.

Yet, the journey through the realm of robotics is not without challenges. Job displacement looms as automation replaces certain tasks, necessitating a reevaluation of the human role in a robotic workforce.

Safety concerns emerge as robots increasingly share spaces with humans, prompting rigorous standards and regulations to ensure secure interactions. Ethical considerations cast a shadow, challenging us to define the boundaries of responsible use and address the potential consequences of unrestrained robotic development. Legal frameworks grapple with the intricacies of liability and accountability in the event of robotic malfunctions or errors. As we stand at this crossroads, the foundational principles of robotic engineering must be accompanied by a conscientious approach, fostering innovation while mitigating risks. The advent of cutting-edge technologies propels the field forward, with artificial intelligence standing as a cornerstone. Machine learning algorithms equip robots with the ability to analyze data, adapt to changing conditions, and improve performance over time. Neural networks, inspired by the human brain, enable robots to recognize patterns, make decisions, and even exhibit a degree of autonomy. The infusion of AI transforms robots from mere tools into intelligent entities capable of learning, reasoning, and problem-solving. This symbiosis of robotics and artificial intelligence heralds a future where machines become not just tools but cognitive companions in our daily lives.

Sensor technologies constitute another frontier in the evolution of robotic engineering. Vision sensors, inspired by human sight, enable robots to perceive and interpret visual information. Tactile sensors, mimicking the sense of touch, allow robots to interact with objects and environments in a nuanced manner. Sensor fusion, the integration of multiple sensor modalities, enhances the depth and accuracy of robotic perception. The mastery of sensors equips robots with a rich understanding of their surroundings, enabling them to navigate dynamic environments and perform complex tasks with precision.

Advanced materials represent a paradigm shift in the physical makeup of robots. Traditional rigid structures give way to soft robotics, inspired by the flexibility and adaptability of natural organisms. Soft robots, constructed from compliant materials, offer unique advantages in terms of safety, dexterity, and interaction with delicate environments. Shape-memory alloys and polymers enable robots to change shape and adapt to different tasks. The exploration of bio-inspired materials allows for the creation of robots that mimic the biomechanics of living organisms. The integration of advanced materials not only expands the capabilities of robots but also broadens the spectrum of potential applications.

The applications of robotics span a diverse array of industries, each reaping the benefits of automation and intelligent systems. In the realm of manufacturing, industrial robots tirelessly perform tasks with precision, efficiency, and speed, revolutionizing production lines. Logistics and warehousing witness the deployment of robots for material handling, inventory management, and order fulfillment, optimizing supply chain operations. The medical field embraces robotic surgery, where the precision of robotic arms enhances the capabilities of surgeons, leading to minimally invasive procedures and faster recovery times.

Beyond the operating room, rehabilitation robotics aids individuals in regaining mobility and independence, offering personalized therapy and support. The exploration of hazardous environments, from the depths of the ocean to outer space, relies on robotic systems that can endure extreme conditions and perform tasks beyond human capabilities. Unmanned aerial vehicles, or drones, navigate the skies for surveillance, mapping, and delivery applications. Service robots find their place in everyday life, from automated vacuum cleaners to interactive social robots in public spaces.

However, the integration of robotics into society raises pertinent questions about the ethical considerations surrounding their use. The displacement of human jobs by automation prompts a critical examination of the societal implications and the need for strategies to address potential unemployment. Safety concerns arise as robots share spaces with humans, necessitating stringent regulations and standards to prevent accidents and ensure secure interactions. Ethical dilemmas emerge regarding the use of robots in sensitive areas, such as healthcare and law enforcement, prompting a reassessment of the boundaries of responsible development and deployment.

Legal frameworks lag behind the rapid advancements in robotics, requiring adaptation to address issues of liability, accountability, and the establishment of guidelines for ethical conduct. As we stand on the cusp of a robotic revolution, it becomes imperative to navigate the ethical landscape with care, ensuring that innovation aligns with human values and societal well-being.

The future of robotics unfolds against the backdrop of emerging trends and innovations that promise to redefine the very nature of human-robot interaction. Swarm robotics, inspired by the collective behavior of social insects, explores the potential of multiple robots working collaboratively to achieve common goals. Soft robotics, with its pliable structures and compliant materials, opens new frontiers in human-robot interaction, enabling robots to interact safely with humans in unstructured environments. The integration of robotics with the Internet of Things (IoT) and the advent of 5G connectivity usher in an era of interconnected devices, where robots seamlessly communicate and collaborate in real-time.

Case studies offer a window into real-world examples of successful robotic implementation, providing insights into the challenges faced, lessons learned, and best practices that pave the way for future endeavors. These narratives illustrate the transformative impact of robotics across diverse industries, shedding light on the collaborative efforts of engineers, researchers, and organizations to push the boundaries of what is possible. As we peer into the future, the speculation and predictions about the trajectory of robotics become a compelling narrative. The emergence of new technologies, the fusion of disciplines, and the integration of robotics into the fabric of daily life create a tapestry of possibilities. The potential societal impact of robotics prompts contemplation about the role of machines in shaping the human experience, raising questions about the ethical, cultural, and economic implications that accompany this technological evolution. In conclusion, the journey through the core principles and concepts of robotic engineering is a voyage into the heart of innovation, where mechanics, electronics, and computer science converge to breathe life into machines. From the early days of industrial automation to the frontiers of artificial intelligence and soft robotics, the evolution of robotics reflects our relentless pursuit of progress and the desire to create machines that transcend the limits of imagination. As we navigate the challenges, explore the applications, and envision the future, robotic engineering stands as a testament to human creativity, pushing the boundaries of what is possible and engineering a future where man and machine coexist in harmony.

Mechanics and Dynamics in Robotics

The field of robotics, a symphony of mechanical, electronic, and computational elements, finds its essential rhythm in the principles of mechanics and dynamics. At the core of robotic engineering lies a profound understanding of how machines move, interact with their surroundings, and execute tasks with precision. Mechanics, the study of the behavior of physical bodies under the influence of forces, forms the foundational framework upon which the mechanics of robots takes shape.

In the intricate dance of robotics, kinematics emerges as a key discipline within mechanics, delving into the geometry of motion. Engineers employ kinematic principles to unravel the intricacies of robotic movement, exploring how joints articulate and how limbs traverse space. The geometric relationships governing these movements become the blueprint for robotic arms that mimic the dexterity of the human hand or for wheeled robots navigating complex terrains. Kinematics, in essence, unlocks the language of motion, allowing engineers to design robots that move precisely and purposefully.

Complementing kinematics, dynamics steps onto the stage, introducing the study of forces and torques that influence the motion of robotic systems. Dynamics provides the narrative behind the scenes, unveiling the unseen forces that propel a robotic arm, guide a drone through the air, or propel a rover across the Martian landscape. Understanding the dynamics of robotic systems is akin to deciphering the hidden choreography of a performance. Engineers apply principles of dynamics to calculate the forces necessary for a robot to lift an object, the torques required for precise movements, and the energy expenditure in executing tasks.

In the world of robotic mechanics, these principles not only dictate the motion of robots but also influence their efficiency, safety, and overall performance. A grasp of mechanics allows engineers to optimize the design of robotic limbs for maximum strength and stability, ensuring that they can handle the rigors of their intended tasks. Dynamics, in turn, guides the development of control systems that manage the intricate balance of forces, enabling robots to move smoothly, respond to external stimuli, and execute tasks with the desired accuracy.

The applications of mechanics and dynamics in robotics extend far beyond the theoretical realm, finding expression in the real-world functionality of robotic systems. In the realm of industrial automation, where precision is paramount, the principles of mechanics and dynamics govern the movement of robotic arms assembling intricate components with micron-level accuracy. These principles also manifest in the realm of robotics-assisted surgery, where the delicacy of human tissues demands an acute understanding of forces to ensure minimal invasiveness and maximal precision.

As we peer into the future, the fusion of mechanics and dynamics with advanced technologies holds the promise of even more sophisticated robotic systems. Soft robotics, inspired by the pliability of natural organisms, introduces new challenges and opportunities in the realm of mechanics. The compliance of soft robotic structures necessitates a reevaluation of traditional rigid-body dynamics, opening avenues for innovative applications in fields such as healthcare and human- robot interaction. As the boundaries of robotics expand, mechanics and dynamics continue to serve as guiding principles, adapting and evolving to meet the demands of an ever-changing technological landscape.

The symbiosis of mechanics and dynamics in robotics is most evident in the development of robotic arms, the quintessential limbs that execute tasks with precision and finesse. The design of robotic arms involves a meticulous consideration of kinematics, determining the range of motion, joint configurations, and overall geometry. Engineers sculpt these mechanical appendages to mimic the dexterity of human limbs or, in some cases, to surpass human capabilities, reaching into spaces too hazardous or remote for human access.
Once the kinematic blueprint is established, dynamics steps into the spotlight, choreographing the forces and torques required for controlled and purposeful movement. The intricate dance between kinematics and dynamics ensures that a robotic arm can lift, manipulate, and place objects with accuracy, replicating or even surpassing the capabilities of the human hand. In industrial settings, robotic arms become the nimble artisans of the assembly line, assembling products with speed, consistency, and precision.
The impact of mechanics and dynamics in robotics extends beyond the manufacturing floor, reaching into fields as diverse as space exploration and healthcare.

Robotic rovers on distant planets navigate the challenging terrain, their movements orchestrated by the principles of mechanics to overcome obstacles and gather valuable data. In the realm of healthcare, robotic surgical systems leverage the precision of mechanics and dynamics to perform minimally invasive procedures, guided by algorithms that ensure precise movements within the human body.

Yet, the realm of mechanics and dynamics in robotics is not confined to rigid structures and articulated joints. The emergence of soft robotics introduces a paradigm shift, where pliable materials and compliant structures redefine the boundaries of movement and interaction. Inspired by the flexibility and adaptability of natural organisms, soft robotics seeks to create machines that can navigate unstructured environments, interact safely with humans, and conform to the complexities of the real world.

Soft robotic systems, constructed from materials with variable stiffness and flexibility, challenge traditional notions of mechanics and dynamics. The compliance of soft structures allows for interactions that go beyond the precision of traditional robotic arms. These robots can gently grip delicate objects, conform to irregular shapes, and adapt to unpredictable scenarios. The principles of dynamics in soft robotics become a delicate orchestration of forces that allow for controlled deformations and movements, offering a new frontier in human-robot collaboration.

In the integration of mechanics and dynamics, control systems emerge as the conductors orchestrating the symphony of robotic movement. Control algorithms, inspired by the principles of feedback and feedforward control, regulate the motors and actuators of a robot to achieve desired motions. Feedback control loops continuously adjust the robot's movements based on sensor data, ensuring that it responds to changes in the environment or deviations from the planned trajectory. Feedforward control anticipates the forces and torques required for specific tasks, optimizing the execution of movements for efficiency and accuracy.

The mastery of control systems is pivotal in the realization of autonomous robots, where the machine navigates and operates without continuous human intervention. In autonomous vehicles, for instance, control systems analyze sensor data to make real-time decisions about steering, acceleration, and braking, adapting to the dynamic nature of road conditions. In the realm of aerial drones, control algorithms stabilize the aircraft, adjust for wind disturbances, and execute precise maneuvers. The intersection of mechanics, dynamics, and control systems creates a synergy that transforms robots from mere tools into intelligent entities capable of adapting to complex and dynamic environments.

As we delve into the intricacies of mechanics and dynamics in robotics, the importance of computational tools becomes apparent. Simulation software allows engineers to model and analyze the behavior of robotic systems before physical prototypes are constructed. These virtual environments provide insights into the dynamics of movement, the impact of forces, and the overall performance of the robot in various scenarios. Computational tools also play a crucial role in the optimization of control algorithms, fine-tuning the parameters for optimal efficiency and responsiveness.

The study of mechanics and dynamics in robotics is a dynamic and ever-evolving discipline. It demands a multidisciplinary approach, requiring expertise in mechanical engineering, electrical engineering, and computer science. Engineers and researchers navigate the complexities of physical systems, seeking a delicate balance between precision and adaptability. As we stand on the precipice of a robotic revolution, the principles of mechanics and dynamics continue to guide us, offering a roadmap for the design, development, and deployment of robotic systems that transcend the boundaries of imagination.

In conclusion, mechanics and dynamics constitute the foundational principles that breathe life into the world of robotics. Kinematics unveils the language of motion, allowing engineers to design robots that move with precision and purpose.

Software and Programming Essentials

In the realm of robotic engineering, where the physical and the digital converge, the mastery of software and programming stands as a linchpin in unlocking the true potential of machines. Beyond the nuts and bolts, the heart of a robot beats in its code, a digital symphony orchestrating movements, decisions, and interactions. Software and programming provide the cognitive essence, transforming robots from mere mechanical entities into intelligent and adaptable systems.

At the core of robotic software lies the symbiotic relationship with computer science, where algorithms and code become the language through which machines perceive, process information, and execute tasks. Artificial intelligence (AI) and machine learning (ML) algorithms stand as the sentinels of robotic cognition, endowing machines with the ability to learn from experience, make decisions, and adapt to dynamic environments. The marriage of software and AI opens avenues for robots to evolve beyond programmed instructions, entering the realm of autonomy where they can navigate complex scenarios with a degree of intelligence.

Programming, the art of crafting instructions for machines, becomes the artisanal touch that shapes the behavior of robots. Engineers use programming languages to imbue machines with the logic and sequence of actions required for specific tasks. The diversity of programming languages, from the lower-level intricacies of C++ to the more abstract and human-readable Python, provides a spectrum of tools for expressing the nuances of robotic behavior. The elegance and efficiency of the code become crucial in optimizing robotic performance, ensuring that the digital soul of the machine aligns seamlessly with its physical capabilities.

In the realm of robotics, there are two broad categories of programming: low-level programming for the control of hardware, and high-level programming for the orchestration of complex behaviors. Low-level programming interfaces directly with the hardware components of a robot, dictating how motors move, sensors collect data, and actuators respond. This level of programming demands precision and an acute understanding of the underlying mechanics. High-level programming, on the other hand, involves crafting algorithms and logic that govern the overarching behavior of the robot. It deals with decision-making processes, path planning, and interactions with the environment.

The integration of software and programming in robotic engineering is perhaps most evident in the development of control systems. These systems serve as the bridge between the digital realm of algorithms and the physical world of mechanical components. Proportional-Integral- Derivative (PID) controllers, a staple in robotics, use feedback mechanisms to regulate the movement of robotic limbs, ensuring that they follow desired trajectories with precision. Control algorithms, crafted through meticulous programming, enable robots to adapt to dynamic environments, respond to unforeseen obstacles, and execute tasks with a level of autonomy.

The development of autonomous robots, capable of operating without continuous human intervention, is a testament to the prowess of software and programming in robotics. Autonomous vehicles, whether on land, in the air, or underwater, rely on sophisticated software algorithms to navigate, avoid obstacles, and make real-time decisions. Path-planning algorithms, a key component of high-level programming, enable robots to find the most efficient routes through complex environments, whether it be a warehouse filled with shelves or a terrain on a distant planet.

In the evolution of robotics, the advent of software frameworks and middleware further accelerates the development process. These frameworks provide a structured environment for engineers to build upon, offering pre-designed modules for common robotic functionalities. ROS (Robot Operating System), for instance, has become a widely adopted open-source framework that simplifies the development of robotic software. Such platforms abstract the complexities of low-level programming, allowing engineers to focus on high-level behaviors and applications.

The significance of software and programming extends beyond the realm of control and autonomy. Computer vision algorithms, an integral part of robotic software, allow machines to interpret visual data and make decisions based on what they "see." Object recognition, image processing, and pattern detection become the tools through which robots interact with their environment. Whether it's a drone avoiding obstacles, a robotic arm identifying objects on an assembly line, or a humanoid robot recognizing human gestures, the marriage of software and computer vision expands the horizons of robotic capabilities.

Machine learning algorithms, a subset of AI, empower robots to learn from data and improve their performance over time. Reinforcement learning, a branch of machine learning, enables robots to learn through trial and error, adapting their behavior based on positive or negative feedback. This capability has profound implications, as robots can refine their actions in response to changing conditions, unexpected challenges, or new tasks.

The integration of software and programming in robotics also addresses the imperative of human-robot interaction. As robots transition from the confines of industrial settings to shared spaces with humans, the ability to understand and respond to human input becomes paramount. Natural Language Processing (NLP) algorithms, part of the software arsenal, enable robots to comprehend and respond to human speech. Gesture recognition algorithms allow machines to interpret human movements, opening avenues for intuitive and non-verbal communication. Human-robot collaboration, a frontier in the evolution of robotics, relies on the finesse of software and programming to ensure seamless interactions and a harmonious coexistence. The interdisciplinary nature of robotic engineering demands a holistic approach to software and programming. The convergence of mechanics, electronics, and software requires engineers to bridge the gap between the physical and the digital seamlessly. This integration becomes particularly evident in the development of robotic interfaces. Human-Machine Interfaces (HMIs) serve as the portals through which humans interact with robots. The design of intuitive and user-friendly interfaces, facilitated by software and programming, becomes crucial for ensuring effective communication and control.

The challenges in the realm of software and programming are as diverse as the applications of robotics themselves. Ensuring the security and reliability of robotic systems demands robust software architecture and rigorous testing protocols. The complexity of high-level programming, especially in the development of autonomous systems, requires a delicate balance between adaptability and predictability. The ethical considerations surrounding AI and machine learning algorithms prompt introspection into the biases that may be inadvertently encoded into robotic behaviors, urging engineers to craft algorithms that align with human values and societal norms.

In conclusion, the role of software and programming in the foundations of robotic engineering is nothing short of transformative. These digital intricacies form the connective tissue between the physical and the cognitive realms of machines, shaping their movements, decisions, and interactions. As we stand on the brink of a robotic revolution, the mastery of software and programming becomes the key to unlocking the full potential of intelligent and autonomous systems. In this dance of algorithms and code, engineers craft the digital soul of robots, ushering in an era where machines become not just tools but adaptive and intelligent companions in our daily lives.

Power and Energy Systems

In the intricate landscape of robotic engineering, where mechanical precision meets computational intelligence, power and energy systems stand as the lifeblood, fueling the movement and functionality of machines. These systems, often overshadowed by the spotlight on mechanics and software, form an indispensable component in the foundations of robotics, dictating the endurance, mobility, and overall efficiency of robotic systems.

At the heart of robotic power and energy systems lies the intricate dance between energy storage, conversion, and distribution. The choice of power source becomes a pivotal decision, influencing the autonomy and versatility of a robot. Batteries, a common power source, store electrical energy for on-demand use. The evolution of battery technologies, from traditional lead-acid to lithium-ion, has ushered in a new era of lightweight, high-energy-density solutions. Beyond batteries, alternative power sources such as fuel cells and solar panels offer unique advantages, especially in scenarios where extended operation or remote deployment is essential.

Efficient energy conversion mechanisms transform stored energy into the mechanical force that propels robots. Motors and actuators, the workhorses of robotic movement, translate electrical energy into precise mechanical actions. The choice of motor type, whether it be DC motors, stepper motors, or servo motors, depends on the specific requirements of the robot. Each motor type brings its own set of advantages, influencing factors such as speed, torque, and accuracy. Efficient energy conversion becomes a balancing act, optimizing the trade-offs between power consumption, performance, and the weight constraints inherent in robotic design.

Power distribution systems form the neural network, ensuring a seamless flow of energy to the various components of a robot. Power distribution boards and electronic circuits manage the allocation of energy, providing the necessary voltage and current to sensors, processors, and actuators. These systems must be designed with precision to minimize energy losses, optimize performance, and ensure the robustness of the entire robotic architecture. As robotic systems become more complex, the design of power distribution systems becomes a critical aspect in maintaining efficiency and reliability.

In the pursuit of autonomous operation, energy efficiency becomes a cornerstone in the design and development of robotic systems. Power consumption directly impacts the endurance of a robot, influencing factors such as mission duration, distance traveled, or the duration of operation between recharging or refueling cycles. Engineers strive to optimize the power-to-weight ratio, seeking the delicate equilibrium between the energy required for movement and the capacity of the power source. Energy-efficient algorithms and control systems become essential tools, enabling robots to perform tasks with minimal energy expenditure and maximum endurance.

The applications of power and energy systems in robotic engineering span a diverse array of industries and scenarios. In industrial automation, where precision and speed are paramount, robotic arms rely on efficient power systems to execute repetitive tasks with minimal downtime. Drones, used for surveillance, mapping, or delivery purposes, navigate the skies powered by lightweight yet high-capacity batteries. Autonomous vehicles, whether on land or underwater, leverage sophisticated power and energy systems to travel extended distances with autonomy.

In the context of space exploration, where the limitations of traditional power sources become apparent, engineers turn to innovative solutions. Solar-powered rovers, such as those deployed on Mars, harness energy from the sun to sustain long-term missions. This reliance on renewable energy sources highlights the adaptability of power and energy systems to diverse environments and challenges. As the boundaries of robotics extend into new frontiers, from the depths of the ocean to the vastness of space, the versatility of power and energy systems becomes a critical factor in enabling exploration and discovery.
The evolution of power and energy systems in robotics is intertwined with advancements in materials and technologies. The development of lightweight materials, capable of storing and conducting energy efficiently, contributes to the overall agility and mobility of robots. Energy-dense batteries and advanced capacitors pave the way for smaller, more powerful robotic systems, expanding the range of applications from medical devices to swarm robotics. Wireless charging technologies offer the promise of continuous operation without the need for physical connections, enhancing the autonomy of robots in dynamic environments.
The challenges inherent in power and energy systems revolve around the quest for sustainability, efficiency, and adaptability. In the age of environmental

consciousness, the ecological footprint of robotic technologies comes under scrutiny. Engineers are tasked with developing energy-efficient solutions that minimize the environmental impact of robotic systems. The recycling and disposal of energy storage components, often laden with complex materials, require thoughtful strategies to mitigate potential environmental hazards. The quest for sustainability prompts the exploration of alternative power sources, such as energy harvesting from ambient sources or the integration of robotics with renewable energy grids.

As robotic systems become more intertwined with our daily lives, addressing the power and energy needs of these machines gains even more significance. Service robots, designed for applications in healthcare, hospitality, and domestic settings, demand power solutions that balance the need for compactness, safety, and longevity. Humanoid robots, developed to interact seamlessly with humans, require power and energy systems that support a wide range of movements and functions. In these contexts, the optimization of power and energy systems becomes not just a technical challenge but a crucial factor in ensuring the acceptance and integration of robots into human-centric environments.

In the context of medical robotics, where precision and reliability are paramount, power and energy systems play a critical role. Surgical robots, designed for minimally invasive procedures, rely on power systems that provide consistent and precise control over robotic arms and instruments. The energy efficiency of these systems becomes crucial in ensuring the safety of patients and the success of medical procedures. Rehabilitation robots, aiding in the recovery of patients, demand power solutions that balance the need for strength with the requirement for gentle and precise movements.

Ethical considerations emerge in tandem with the advancements in power and energy systems. The potential misuse of robotic technologies, especially in the realm of autonomous weapons or surveillance systems, prompts a reevaluation of the ethical implications of providing machines with advanced power capabilities. The responsible and ethical deployment of robots, guided by frameworks that ensure transparency, accountability, and adherence to societal values, becomes imperative as power and energy systems evolve.

In conclusion, power and energy systems stand as the silent architects of robotic functionality, providing the essential life force that empowers machines to move, perceive, and interact. The evolution of these systems reflects the relentless pursuit of efficiency, sustainability, and adaptability in the field of robotic engineering. As we peer into the future, where robots become integral companions in our daily lives, the role of power and energy systems becomes not just a technical challenge but a critical factor in shaping the coexistence of humans and machines. In this dynamic landscape, the fusion of mechanical prowess with computational intelligence, powered by advanced and sustainable energy solutions, paves the way for a future where robots seamlessly integrate into the fabric of our existence.

CHAPTER II

Innovations in Robotic Hardware

Advanced Sensors and Perception Systems

In the ever-evolving realm of robotic engineering, the pursuit of heightened awareness and perceptive capabilities has given rise to a new era marked by advanced sensors and perception systems. These innovations in robotic hardware transcend the traditional boundaries of mechanical precision, endowing machines with the ability to sense, interpret, and respond to the complexities of the surrounding environment. In this symbiotic dance between hardware and intelligence, robots transition from mere tools to sophisticated entities that can navigate dynamic scenarios, interact with humans, and adapt to unforeseen challenges.

At the forefront of this revolution are advanced sensor technologies, each designed to cater to specific aspects of the robot's interaction with the world. Vision sensors, inspired by the intricacies of human sight, are pivotal in the realm of robotics. Cameras equipped with sophisticated image processing algorithms enable robots to "see" and interpret visual data. From object recognition to depth perception, vision sensors lay the foundation for a robot's understanding of its surroundings. This ability to visually interpret the environment allows robots to navigate unstructured spaces, identify objects, and even interact with humans in a more intuitive manner.

Beyond the visual spectrum, robots leverage a diverse array of sensors to gather information about their surroundings. LiDAR (Light Detection and Ranging) sensors employ laser beams to measure distances and create detailed 3D maps of the environment. This technology is instrumental in robotic navigation, especially in scenarios where precise mapping is essential, such as in autonomous vehicles or drones. Similarly, radar sensors utilize radio waves to detect objects and assess their speed and direction. These sensors find applications in robotics ranging from collision avoidance systems to monitoring the movement of objects in industrial settings.

Tactile sensors, inspired by the sense of touch, bring a new dimension to robotic interaction. These sensors, often integrated into robotic grippers or limbs, allow machines to perceive force, pressure, and texture. The sense of touch becomes a powerful tool in delicate tasks, such as handling fragile objects or collaborating with humans in shared workspaces. The integration of tactile sensors enhances the safety and precision of robots, enabling them to interact with the physical world in a nuanced manner.

The fusion of these sensor technologies forms the bedrock of perception systems in robotics. Perception systems go beyond the mere collection of data; they involve the interpretation and understanding of the information gathered by sensors. Machine learning algorithms, a subset of artificial intelligence, play a pivotal role in perception systems, enabling robots to learn from data, recognize patterns, and make informed decisions. This cognitive aspect elevates robotic hardware from reactive machines to entities capable of adaptive and intelligent responses.

One of the groundbreaking innovations in perception systems is the advent of computer vision. This interdisciplinary field integrates computer science and vision sensors, allowing robots to interpret visual data in ways that mimic human vision. Object detection, image recognition, and facial expression analysis become routine tasks for robots equipped with advanced computer vision capabilities. This innovation extends the applications of robotics into diverse fields, from surveillance and security to healthcare and autonomous vehicles.

The application of advanced sensors and perception systems extends into the realm of human-robot interaction, where the ability to understand and respond to human cues becomes paramount. Speech recognition sensors enable robots to comprehend spoken commands, facilitating natural language interaction. Gesture recognition technologies interpret human movements, allowing robots to respond to non-verbal cues. This level of interaction, inspired by human communication modalities, enhances the usability and acceptance of robots in various contexts, from customer service to assistive devices for individuals with disabilities.

In healthcare, robots equipped with advanced sensors and perception systems are revolutionizing patient care. Surgical robots, guided by computer vision and haptic feedback systems, assist surgeons in performing minimally invasive procedures with unprecedented precision. Wearable robots, fitted with sensors that monitor physiological parameters, provide real-time health data for patients and healthcare professionals. Social robots, designed to assist and engage with individuals in healthcare settings, leverage advanced perception systems to understand and respond to human emotions and needs.

The integration of advanced sensors and perception systems has a profound impact on the field of autonomous vehicles. In self-driving cars, a fusion of vision sensors, LiDAR, radar, and sophisticated perception algorithms enables the vehicle to navigate complex traffic scenarios, interpret road signs, and respond to unexpected obstacles. These innovations hold the promise of revolutionizing transportation, reducing accidents, and enhancing the efficiency of urban mobility.

The synergy between advanced sensors and perception systems becomes particularly evident in the realm of industrial robotics. Collaborative robots, or cobots, work alongside humans in shared workspaces, necessitating a high degree of awareness and adaptability. Vision sensors enable cobots to recognize and respond to human presence, ensuring a safe working environment. Tactile sensors, integrated into robotic arms, allow for precise interaction with delicate materials or intricate tasks. Perception systems in industrial robots go beyond mere automation; they enable machines to collaborate seamlessly with human workers, opening new frontiers in efficiency and flexibility.

While the integration of advanced sensors and perception systems propels robotics into new realms of capability, it also presents challenges. The sheer volume of data generated by sensors necessitates sophisticated processing capabilities, placing demands on computational resources. The need for real-time decision-making in dynamic environments requires algorithms that balance accuracy with speed. Ethical considerations surrounding privacy and data security come to the forefront as robots equipped with advanced perception systems become more prevalent in public spaces and private settings.

In the context of artificial intelligence, the use of machine learning algorithms for perception introduces complexities related to bias and interpretability. The algorithms "learn" from the data they are exposed to, and if the training data contains biases, the system may perpetuate and amplify these biases in its decision-making. As robots become more integrated into societal functions, addressing these ethical considerations becomes crucial to ensuring fair and responsible use of advanced sensors and perception systems.

Actuators and Mechanisms

In the intricate tapestry of robotic engineering, actuators and mechanisms emerge as the unsung heroes, orchestrating the ballet of movement that defines the essence of robots. These components constitute the very sinews and joints, the mechanical muscles and bones that transform the theoretical potential of a machine into tangible and purposeful actions. In the evolution of robotic hardware, innovations in actuators and mechanisms not only enhance the precision and agility of machines but also broaden the spectrum of applications, ushering in an era where robots move with finesse, adaptability, and a degree of biomechanical artistry.

At the heart of robotic movement lies the actuator, the mechanical device responsible for converting energy into motion. Actuators serve as the powerhouse, propelling the limbs, joints, and various components of a robot with the force required to execute tasks. The diversity of actuators mirrors the versatility demanded by the myriad applications of robotics. Electric motors, including DC motors, stepper motors, and servo motors, dominate the landscape, providing the torque and precision necessary for controlled movements. Hydraulic and pneumatic actuators, harnessing the power of fluids, find their place in scenarios requiring high force and rapid response, such as industrial automation and heavy machinery.

Stepping beyond conventional actuators, innovations in this realm herald the advent of smart and adaptive technologies. Shape-memory alloys, for instance, exhibit the unique ability to change shape in response to temperature variations, allowing for dynamic and programmable movements. Piezoelectric actuators leverage the property of certain materials to deform under an electric field, enabling precise and rapid motion in microscale applications. These innovations not only expand the repertoire of movements but also contribute to the miniaturization and intricacy of robotic systems, paving the way for applications in fields as diverse as medical robotics and micro-manipulation.

Mechanisms, the mechanical systems that govern the motion and interaction of robotic components, complement actuators in the symphony of movement. The choice of mechanisms dictates the range of motion, stability, and overall functionality of a robot. Robotic arms, inspired by the kinematics of the human limb, leverage articulated mechanisms with joints that mimic the flexibility and dexterity of the human arm. Parallel mechanisms, where multiple actuators work in parallel to control a single end effector, offer enhanced stability and precision, making them suitable for applications requiring meticulous control, such as surgical robots.

In the realm of locomotion, mechanisms for wheeled and tracked robots optimize movement across diverse terrains. Legged mechanisms, inspired by the biomechanics of animals, introduce a dynamic dimension to robotic mobility. Bipedal and quadrupedal robots, equipped with articulated legs, navigate complex environments with agility and adaptability. The study of biomimicry, emulating the movements and structures found in nature, inspires innovations in robotic mechanisms that transcend the traditional constraints of wheeled or tracked locomotion.

Soft robotics, an emerging frontier, redefines the very nature of robotic mechanisms. Inspired by the pliability and compliance of natural organisms, soft robots utilize flexible materials and deformable structures. Pneumatic and hydraulic systems drive soft robotic movements, allowing for safe and gentle interactions with delicate environments, as well as humans. The integration of soft robotic mechanisms expands the possibilities in fields like medical applications, where the gentle touch of soft actuators proves beneficial in delicate procedures and rehabilitation.

The synergy of actuators and mechanisms extends into the intricacies of haptics, introducing a tactile dimension to robotic interaction. Haptic mechanisms enable robots to sense and respond to touch, providing feedback to users or adapting to variations in the environment. This tactile proficiency becomes instrumental in applications such as teleoperation, where a human operator remotely controls a robot and relies on haptic feedback to feel and respond to the surroundings. The development of haptic mechanisms opens avenues for immersive experiences in virtual reality, where users can feel and interact with virtual objects through robotic interfaces.

As the boundaries of robotic hardware expand, the integration of actuators and mechanisms becomes pivotal in the development of robotic exoskeletons. These wearable robotic systems, equipped with actuators and mechanisms that augment human movement, find applications in medical rehabilitation, physical assistance, and industrial ergonomics. Powered by electric, pneumatic, or hydraulic actuators, exoskeletons assist users in tasks requiring strength and endurance, offering a bridge between human capabilities and robotic augmentation.

The evolution of actuators and mechanisms is deeply entwined with the pursuit of energy efficiency and sustainability. Innovations in materials, such as lightweight alloys and composites, contribute to the development of energy-efficient actuators that reduce power consumption without compromising performance. Additionally, the exploration of energy recovery mechanisms, where the motion of a robot is used to generate and store energy, aligns with the broader drive toward sustainable robotics. These advancements not only enhance the autonomy of robots but also address the environmental impact of their operations.

In the intricate dance of actuators and mechanisms, control systems emerge as the conductors orchestrating the symphony of robotic movement. Control algorithms, ranging from simple proportional-integral-derivative (PID) controllers to advanced adaptive control systems, regulate the motion of actuators and mechanisms with precision. Feedback control loops continuously adjust the robot's movements based on sensor data, ensuring that it responds to changes in the environment or deviations from the planned trajectory. The mastery of control systems becomes pivotal in the realization of autonomous robots, where the machine navigates and operates without continuous human intervention.

The integration of actuators and mechanisms extends beyond the realms of industrial automation and manufacturing, reaching into diverse fields that impact our daily lives. In the context of service robots, designed for tasks in hospitality, healthcare, or customer service, the proficiency of actuators and mechanisms determines the effectiveness and reliability of tasks performed. From delivering goods in warehouses to assisting patients in hospitals, service robots equipped with precise and adaptive movement capabilities become valuable assets in various sectors.

In the exploration of hazardous or inaccessible environments, robotic mechanisms and actuators play a crucial role. Robotic arms on planetary rovers, powered by intricate actuators, manipulate scientific instruments to analyze the surface of distant celestial bodies. Underwater robots, equipped with propulsion mechanisms and manipulators, delve into the depths of the ocean to study marine ecosystems or perform maintenance on underwater infrastructure. In these scenarios, the robustness and adaptability of robotic hardware become essential for the success of missions in harsh and remote environments.

Challenges in the realm of actuators and mechanisms revolve around the delicate balance between performance and complexity. The quest for lightweight and compact actuators that deliver high torque or force without sacrificing efficiency is an ongoing pursuit. The intricacies of designing mechanisms that offer both stability and agility, especially in complex robotic systems, demand a deep understanding of kinematics and dynamics. Control algorithms face the challenge of adapting to diverse scenarios and uncertainties, requiring a nuanced approach to achieve the desired balance between autonomy and responsiveness.

In conclusion, the innovations in actuators and mechanisms represent a pivotal chapter in the narrative of robotic hardware. These components, often hidden beneath the surface, breathe life into machines, endowing them with the ability to move, interact, and adapt. As robotics continues to evolve, the artistry of movement facilitated by advanced actuators and mechanisms expands the horizons of what is possible. From the intricacies of surgical robots to the adaptability of soft robotic systems, the integration of these hardware innovations shapes a future where robots seamlessly integrate into our lives, offering not just functionality but a graceful and purposeful presence.

Materials Revolutionizing Robotics

In the dynamic realm of robotics, the material sciences play a pivotal role, serving as the silent architects that define the form, function, and capabilities of machines. The choice of materials in robotic design is more than a matter of structural integrity; it is a strategic decision that shapes the performance, adaptability, and even the aesthetic appeal of robots. In the ongoing quest for innovation, a materials revolution is underway, ushering in a new era where smart materials, composites, and nanotechnologies redefine the boundaries of what robots can achieve. This revolution transcends the traditional constraints of metal and plastic, unleashing a wave of possibilities that elevate robotics to unprecedented heights.

At the forefront of this revolution are smart materials, a class of materials endowed with dynamic properties that respond to external stimuli. Shape-memory alloys, for instance, have the ability to return to a predefined shape when subjected to heat or stress. This unique characteristic finds applications in robotics, allowing for the creation of self-reconfigurable structures or shape-shifting components. The integration of shape-memory alloys into robotic systems introduces a level of adaptability and flexibility that was once the realm of science fiction. Robots with the ability to change their form based on environmental conditions open avenues for applications in diverse fields, from medical devices to space exploration.

Piezoelectric materials represent another category of smart materials that are making waves in the robotics landscape. These materials generate an electric charge in response to mechanical stress, and conversely, they can deform when subjected to an electric field. In robotics, piezoelectric actuators and sensors enable precise and rapid movements, contributing to the development of miniature and highly responsive robotic systems. The application of piezoelectric materials extends into areas such as haptics, where the sense of touch is enhanced through the integration of these materials into robotic interfaces. As robots become more intertwined with human-centric applications, the use of piezoelectric materials enhances the sophistication of tactile interactions.

The advent of soft robotics, a subfield propelled by advancements in flexible and deformable materials, is transforming our understanding of robotic design. Soft robots mimic the pliability and compliance found in natural organisms, allowing for gentle interactions with the environment and humans. Elastomers, hydrogels, and other soft materials replace rigid components, enabling robots to navigate complex and unstructured spaces with ease. Soft robotic grippers, inspired by the dexterity of an octopus tentacle, adapt to the shape of objects, offering a versatile approach to manipulation. This innovation in materials not only broadens the scope of robotic applications but also addresses safety concerns in scenarios where robots collaborate with humans.

Composites, engineered materials composed of two or more constituent materials with distinct properties, stand as another cornerstone in the materials revolution. Fiber-reinforced composites, for instance, combine the strength of fibers with the lightweight properties of a matrix material. In robotics, this translates into components that are both robust and lightweight, a crucial combination for applications in aerospace, automotive, and industrial robotics. The integration of composites into robotic structures enhances efficiency without compromising structural integrity. Carbon fiber composites, known for their exceptional strength-to-weight ratio, find applications in the construction of robotic limbs, frames, and even entire exoskeletons, contributing to the development of agile and high-performance robots.

Nanomaterials, at the forefront of cutting-edge research, are redefining the very building blocks of robotic components. Carbon nanotubes and graphene, with their extraordinary electrical, thermal, and mechanical properties, hold the promise of revolutionizing various aspects of robotic design. These nanomaterials contribute to the development of lightweight yet incredibly strong components, enhancing the overall performance of robots. In the realm of sensors, nanomaterials enable the creation of highly sensitive and responsive devices, expanding the capabilities of robots in perception and interaction. The integration of nanomaterials into robotic systems not only enhances their functionality but also paves the way for the miniaturization of components, a crucial factor in the development of micro- and nanorobotics.

Biocompatible materials, designed to interact seamlessly with biological systems, bridge the gap between robotics and the human body. In medical robotics, the use of biocompatible materials in implants and prosthetics ensures compatibility and reduces the risk of adverse reactions. Soft and flexible materials, mimicking the properties of human tissues, enable the development of robotic devices that can operate within the body, such as soft robotic endoscopes or minimally invasive surgical robots. The integration of biocompatible materials into robotic design aligns with the growing field of biohybrid robotics, where machines interact with and augment biological systems.

The materials revolution in robotics extends its influence into the field of energy storage and harvesting. Advanced materials contribute to the development of lightweight and high-capacity batteries, essential for the autonomy of robots. Graphene-based supercapacitors, for example, offer rapid charging and discharging capabilities, addressing one of the key challenges in robotics—power supply. The integration of energy-harvesting materials, such as piezoelectric or thermoelectric materials, allows robots to generate power from their environment, reducing dependence on traditional power sources. This aspect becomes crucial in scenarios where robots operate in remote or inaccessible locations.

In the quest for sustainability, bio-based and recyclable materials emerge as pivotal players in the materials revolution. The use of biomimetic materials, inspired by natural structures and processes, contributes to the development of lightweight and resilient robotic components. The exploration of bio-inspired materials, such as synthetic polymers inspired by spider silk or adhesives inspired by gecko feet, leads to the creation of robots that emulate the efficiency and adaptability found in nature. Additionally, the integration of recyclable materials aligns with the broader commitment to eco-friendly robotics, ensuring that the environmental impact of robotic technologies is minimized.

Challenges in the materials revolution revolve around the delicate balance between innovation and practicality. The integration of novel materials often comes with the need for extensive testing and validation to ensure reliability and safety. The scalability of production processes for advanced materials poses challenges in meeting the demands of mass-produced robotic systems. Additionally, the economic considerations associated with the cost of advanced materials must be addressed to make these innovations accessible for a broader range of applications.

In conclusion, the materials revolution in robotics represents a paradigm shift that transcends the traditional boundaries of robotic design. Smart materials, composites, nanotechnologies, and bio-inspired materials collectively propel the field into uncharted territories of innovation. As robots evolve from rigid and mechanistic entities to adaptive, flexible, and biomimetic systems, the materials they are constructed from become the bedrock of this transformation. The ongoing exploration of novel materials not only expands the horizons of robotic capabilities but also fosters a new era where machines seamlessly integrate into the fabric of our daily lives, offering not just functionality but a harmonious and intelligent presence.

The Rise of Modular Robotics

In the ever-evolving landscape of robotics, a transformative wave is reshaping the very fabric of how we conceptualize and construct machines. At the forefront of this revolution is the rise of modular robotics—an approach that redefines the traditional notions of robotic design and functionality. Modular robotics depart from the monolithic, single-entity machines of the past, introducing a paradigm where robots are constructed from interchangeable, independent modules. This departure from the conventional not only reimagines the possibilities of robotic systems but also ushers in an era of adaptability, scalability, and a level of customization that aligns with the diverse demands of modern applications.

The essence of modular robotics lies in the concept of modularity—a design philosophy where a complex system is broken down into discrete, interchangeable modules that can be combined and recombined to create various configurations. This departure from a singular, fixed form introduces a level of flexibility that addresses the challenges posed by dynamic and evolving environments. Each module, in itself, is a self-contained unit with its own set of functionalities, creating a plug- and-play ecosystem where modules seamlessly integrate to perform tasks beyond the capability of a single unit.

One of the primary advantages of modular robotics is the inherent adaptability it offers. Traditional robots are often designed for specific tasks or environments, limiting their versatility. In contrast, modular robots, equipped with a repertoire of specialized modules, can be easily reconfigured to suit different tasks or navigate varied terrains. For instance, a modular robot designed for exploration could be reassembled with modules optimized for manipulation, transforming it into a versatile robotic arm. This adaptability is particularly advantageous in scenarios where tasks are unpredictable, or where robots need to operate in diverse and dynamic environments, such as disaster response or search and rescue missions.

Scalability is another hallmark of modular robotics. The ability to add or remove modules allows for the creation of robots that can scale in size or capability based on the requirements of a specific task. Small-scale modular robots can collaborate to form larger structures, adapting to the scope and demands of a mission. This scalability feature is particularly relevant in scenarios where the size or complexity of tasks varies, such as in industrial settings where robots may need to handle different-sized objects or in agricultural applications where robotic systems must navigate varying field sizes.

The versatility of modular robotics extends to the realm of customization. Instead of designing a new robot for each unique task, engineers can assemble bespoke robots tailored to specific applications by selecting and combining the appropriate modules. This not only streamlines the design and development process but also reduces costs associated with creating specialized robots for each task. Customization becomes a powerful tool in addressing the diverse needs of industries, from manufacturing and logistics to healthcare and research.

The emergence of modular robotics has been propelled by advancements in materials, communication technologies, and control systems. Lightweight materials contribute to the development of modules that are easy to handle and integrate, while robust communication interfaces enable seamless interaction between modules. Control systems, often centralized or distributed across modules, orchestrate the collaborative efforts of the individual units, ensuring cohesive and synchronized behavior. The integration of sensors into modules enhances the perception capabilities of modular robots, allowing them to adapt to changes in the environment or collaborate with precision.

Swarm robotics, a subset of modular robotics, exemplifies the collective power of modular systems. In swarm robotics, a multitude of simple, often identical, modules collaborate to perform tasks with collective intelligence. Inspired by the collective behavior observed in social insects, such as ants or bees, swarm robotics leverages the strength of numbers and distributed decision-making to achieve complex objectives. Swarm robots excel in applications such as environmental monitoring, where a large number of distributed modules can cover vast areas with efficiency, or in disaster scenarios, where swarm robots can collaborate in search and rescue missions.

The application of modular robotics is diverse and spans a multitude of sectors. In manufacturing, modular robotic systems can adapt to changing production demands by reconfiguring their modules to handle different tasks or assemble various products. The flexibility of modular robots makes them well-suited for tasks in unstructured environments, such as construction sites or agricultural fields, where adaptability to diverse conditions is paramount. In logistics and warehousing, modular robots can collaborate to efficiently manage inventory, handle packages of varying sizes, and navigate complex warehouse layouts.

The healthcare industry stands to benefit significantly from the rise of modular robotics. Surgical robots, composed of modular units equipped with specialized tools, can be customized for specific procedures, enhancing precision and minimizing invasiveness. Rehabilitation robots, designed to assist individuals in regaining mobility or strength, can be adapted to the specific needs of patients through the reconfiguration of modules. The modular approach allows for the creation of robotic exoskeletons tailored to individual rehabilitation plans.

The rise of modular robotics has implications for education and research, providing a platform for hands-on learning and experimentation. Educational robotics kits, based on modular principles, allow students to build and program their robots, gaining practical insights into robotics and engineering concepts. Researchers benefit from the flexibility of modular systems, enabling them to prototype and test various robotic configurations without the need for extensive redesigns.

However, challenges accompany the rise of modular robotics. The coordination and communication between modules, especially in large-scale systems, pose significant computational and algorithmic challenges. Ensuring robustness and reliability in the face of module failures or changes in the environment requires sophisticated control strategies. Standardization of interfaces and communication protocols is an ongoing consideration to enhance interoperability and compatibility between modules from different manufacturers.

Ethical considerations also come into play, particularly in the context of swarm robotics. The potential societal impact of large numbers of autonomous modules collaborating raises questions about privacy, security, and the ethical use of these systems. As modular robotics becomes more integrated into daily life, the responsible development and deployment of these systems necessitate careful consideration of ethical implications and the establishment of guidelines for their ethical use.

In conclusion, the rise of modular robotics signifies a paradigm shift in the field, ushering in a new era where robots are not fixed entities but dynamic, adaptable systems. The modularity approach aligns with the ever-changing and diverse needs of modern applications, offering a flexible and scalable solution to the challenges posed by dynamic environments. As modular robotics continues to evolve, it holds the promise of transforming industries, enhancing human-robot collaboration, and contributing to the development of intelligent systems capable of navigating the complexities of the world. This paradigm shift not only shapes the future of robotics but also invites us to rethink the very nature of machines and their role in our interconnected and dynamic world.

CHAPTER III

Artificial Intelligence in Robotics

AI Integration in Robotics

In the intricate tapestry of technological progress, the convergence of artificial intelligence (AI) and robotics stands as a testament to the transformative power of synergistic innovation. The integration of AI in robotics represents a paradigm shift, transcending the traditional boundaries of automated machines and ushering in a new era where robots are not just programmable tools but intelligent entities capable of learning, adapting, and interacting with the world in increasingly sophisticated ways. This symbiotic relationship between AI and robotics is reshaping industries, revolutionizing the nature of work, and offering a glimpse into a future where machines seamlessly collaborate with humans.

At the heart of AI integration in robotics lies the concept of autonomy. Traditionally, robots were programmed with explicit instructions to perform specific tasks within predefined environments. The infusion of AI introduces a level of autonomy that allows robots to perceive, interpret, and respond to dynamic and unstructured scenarios. Machine learning algorithms, a subset of AI, enable robots to learn from data, recognize patterns, and make decisions without explicit programming. This adaptability is particularly crucial in environments where tasks are complex, environments are dynamic, or where robots need to operate alongside humans.

The marriage of AI and robotics becomes evident in the realm of perception and sensor technologies. AI algorithms, particularly those in the domain of computer vision, empower robots with the ability to interpret visual data much like humans. Cameras equipped with sophisticated image recognition algorithms enable robots to "see" and understand their surroundings. Object recognition, facial detection, and scene understanding become routine tasks for robots, allowing them to navigate and interact with the world in a manner that transcends traditional programming constraints. The integration of AI-enhanced perception systems extends the application of robotics into diverse fields, from autonomous vehicles to surveillance and security.

Beyond vision, AI contributes to advancements in sensor technologies that enable robots to sense and interact with the physical world. Tactile sensors, capable of detecting pressure, force, and texture, imbue robots with a sense of touch, facilitating delicate interactions and enhancing their ability to manipulate objects. The fusion of AI with sensor technologies creates a feedback loop where robots can continuously adapt their movements based on real-time information, enabling precise and responsive actions. This level of sensorimotor integration is particularly valuable in applications such as industrial automation, where robots collaborate with human workers or handle fragile materials.

In the realm of decision-making, AI plays a pivotal role in transforming robots into intelligent entities capable of navigating complex scenarios. Planning and control algorithms, driven by AI techniques, enable robots to optimize their movements, strategize tasks, and make decisions based on a dynamic understanding of the environment. This cognitive aspect of robotics elevates machines from mere executants of preprogrammed instructions to entities capable of adaptive and intelligent responses. In autonomous vehicles, for example, AI algorithms analyze sensor data to make split-second decisions, navigate traffic, and respond to unforeseen obstacles.

Machine learning, a cornerstone of AI, introduces the concept of learning from experience into the realm of robotics. Reinforcement learning, a type of machine learning, allows robots to learn optimal behavior through trial and error. Robots equipped with reinforcement learning algorithms can refine their actions based on feedback from the environment, gradually improving their performance over time. This iterative learning process is invaluable in scenarios where tasks are complex, and the optimal strategy may not be explicitly known. In industrial settings, robots employing reinforcement learning can adapt to variations in manufacturing processes, optimizing efficiency and minimizing errors.

The integration of AI in robotics extends into the domain of natural language processing and human-robot interaction. Speech recognition algorithms empower robots with the ability to comprehend and respond to spoken commands, facilitating intuitive communication. This is particularly relevant in applications where robots collaborate with humans, such as in customer service or healthcare. Additionally, AI-driven gesture recognition technologies enable robots to interpret non-verbal cues, enhancing their ability to understand and respond to human gestures. The fusion of these capabilities creates a more natural and user-friendly interaction between humans and robots.

The field of collaborative robotics, often referred to as cobots, exemplifies the harmonious integration of AI and robotics in shared workspaces. Cobots are designed to work alongside humans, leveraging AI algorithms to adapt their movements and behavior in response to human presence and actions. Advanced safety features, enabled by AI, ensure that cobots can operate safely in proximity to human workers. The collaborative nature of these robots opens new frontiers in industrial automation, where humans and robots can work together seamlessly, each leveraging their unique strengths.

In healthcare, AI integration in robotics holds the promise of transforming patient care. Surgical robots, guided by AI algorithms, enhance the precision and capabilities of surgeons in minimally invasive procedures. AI-driven diagnostic robots analyze medical data, assist in the identification of diseases, and contribute to personalized treatment plans. Social robots, designed to interact with patients in healthcare settings, leverage AI for understanding and responding to human emotions, offering companionship and support. The synergy between AI and robotics in healthcare not only improves the efficiency of medical procedures but also enhances the overall patient experience.

The advent of swarm robotics, a collaborative approach inspired by the collective behavior observed in nature, relies heavily on AI for decentralized decision-making. Swarms of small robots, communicating with each other through AI algorithms, can achieve complex tasks collectively. This approach finds applications in scenarios such as environmental monitoring, where a large number of distributed robots can cover vast areas efficiently, or in disaster response, where swarms of robots collaborate in search and rescue missions. The decentralized intelligence of swarm robotics exemplifies how AI can amplify the capabilities of robotic systems through collaboration.

Despite the transformative potential of AI integration in robotics, challenges abound. The ethical considerations surrounding the deployment of intelligent robots raise questions about accountability, transparency, and bias. Ensuring that AI algorithms operate ethically and without perpetuating societal biases is a critical concern. The explainability of AI decisions becomes crucial, particularly in scenarios where robots operate in sensitive domains such as healthcare or criminal justice. Striking a balance between autonomy and human oversight is an ongoing challenge, requiring careful consideration of the ethical implications of AI-driven robotic systems.

Technical challenges include the need for robustness and adaptability in AI algorithms. Real-world environments are dynamic, unpredictable, and often involve uncertainties that can challenge the reliability of AI-driven robotic systems. Ensuring that robots can adapt to changes in the environment, handle unforeseen obstacles, and continue to operate safely is a continuous area of research. The optimization of computational efficiency is also critical, especially in scenarios where robots operate in real-time or with limited computational resources.

The integration of AI in robotics contributes to ongoing debates about the impact of automation on employment. While AI-driven robotics can enhance productivity and efficiency, concerns about job displacement arise as routine tasks become automated. The ethical and societal implications of these changes necessitate careful consideration and the exploration of solutions that balance the benefits of automation with the preservation of employment opportunities.

In conclusion, the integration of AI in robotics marks a transformative juncture in the evolution of intelligent machines. The marriage of automation and intelligence brings forth a new generation of robots that not only execute tasks with precision but also learn, adapt, and collaborate with humans. This symbiotic relationship between AI and robotics holds the promise of revolutionizing industries, enhancing healthcare, and redefining the nature of work. As AI-driven robots become increasingly integrated into our daily lives, the responsible development and ethical deployment of these technologies become imperative to ensure a future where machines and humans coexist harmoniously, leveraging the strengths of each for the betterment of society.

Machine Learning and Robotics

In the dynamic intersection of machine learning and robotics, a profound synergy is reshaping the landscape of automation, heralding an era where machines not only execute predefined tasks but learn and adapt in response to their environments. The fusion of these two transformative technologies empowers robots with a level of intelligence and autonomy that transcends traditional programming paradigms, opening doors to unprecedented possibilities in various domains.
At the core of this integration lies the paradigm of machine learning—a subset of artificial intelligence (AI)

that endows machines with the ability to learn from data and improve their performance over time without being explicitly programmed. When applied to robotics, machine learning algorithms become the cognitive engine that elevates robots from mere executors of preprogrammed instructions to adaptable, intelligent entities capable of making decisions, recognizing patterns, and evolving based on experience.

Perception, a fundamental aspect of robotic interaction with the environment, is revolutionized by machine learning. Computer vision algorithms, powered by machine learning, enable robots to interpret visual data with a sophistication that rivals human perception. Cameras equipped with these algorithms allow robots to "see" and understand the world, recognizing objects, people, and spatial relationships. This transformative capability finds applications across diverse fields, from autonomous vehicles navigating complex traffic scenarios to industrial robots identifying and manipulating objects in unstructured environments. The marriage of machine learning and robotic perception extends beyond vision to include other senses. Tactile sensors, equipped with machine learning algorithms, enable robots to feel and interact with their surroundings. These sensors can detect textures, pressures, and forces, providing robots with a sense of touch that facilitates delicate and precise manipulation. The integration of tactile machine learning enhances the dexterity of robotic hands, making them capable of handling fragile objects or adapting to variations in materials.

In the realm of decision-making, machine learning introduces a dynamic and adaptive intelligence to robotics. Planning and control algorithms, imbued with machine learning techniques, allow robots to optimize their movements, strategize tasks, and make decisions in complex and dynamic environments. Reinforcement learning, a type of machine learning, enables robots to learn optimal behavior through trial and error, adapting their actions based on feedback from the environment. This learning from experience is particularly valuable in scenarios where tasks are intricate and the optimal strategy may not be explicitly known.

Machine learning algorithms extend their influence into human-robot interaction, enhancing the ability of robots to understand and respond to human commands. Speech recognition, a machine learning application, enables robots to comprehend spoken language, facilitating intuitive communication. Robots equipped with natural language processing algorithms can understand and respond to verbal instructions, making them more accessible and user-friendly. This capability is particularly relevant in applications where robots collaborate with human users, such as in customer service or smart home environments.

Collaborative robotics, a field that emphasizes the coexistence of humans and robots in shared workspaces, leverages machine learning to enhance safety and adaptability. Machine learning algorithms enable robots to recognize and respond to human presence, ensuring safe and efficient collaboration. These collaborative robots, often referred to as cobots, can adapt their movements and behaviors based on the actions of human coworkers, creating a harmonious and productive working environment.

The concept of autonomy in robotics is redefined by machine learning. Traditionally, robots were programmed with explicit instructions to perform specific tasks within predefined environments. Machine learning algorithms introduce a level of autonomy that allows robots to learn, adapt, and make decisions in real-time. This adaptability is particularly crucial in environments where tasks are complex, environments are dynamic, or where robots need to operate alongside humans.

Swarm robotics, a collaborative approach inspired by the collective behavior observed in nature, relies heavily on machine learning for decentralized decision-making. Swarms of small robots, communicating with each other through machine learning algorithms, can achieve complex tasks collectively. This decentralized intelligence, facilitated by machine learning, allows swarms of robots to exhibit emergent behaviors and adapt to changing conditions. Applications of swarm robotics range from environmental monitoring to disaster response, where large numbers of distributed robots can cover vast areas with efficiency.

In healthcare, the integration of machine learning and robotics holds significant promise. Surgical robots, guided by machine learning algorithms, enhance the precision and capabilities of surgeons in minimally invasive procedures. Machine learning algorithms analyze medical data, assist in the identification of diseases, and contribute to personalized treatment plans. Social robots, designed to interact with patients in healthcare settings, leverage machine learning for understanding and responding to human emotions, offering companionship and support.

The transformative impact of machine learning and robotics extends into education and research. Educational robotics kits, based on machine learning principles, provide students with hands-on experiences in building and programming robots. Researchers benefit from the adaptability and learning capabilities of machine learning algorithms, enabling them to prototype and test various robotic configurations without the need for extensive reprogramming.

Challenges accompany the integration of machine learning and robotics, ranging from technical hurdles to ethical considerations. The robustness and reliability of machine learning algorithms in real-world environments, where uncertainties and variations are inherent, pose ongoing challenges. Ensuring the explainability and transparency of AI-driven robotic decisions becomes crucial, especially in sensitive domains such as healthcare or criminal justice. Ethical considerations, including bias in machine learning algorithms and the impact on employment, necessitate careful scrutiny and the establishment of guidelines for responsible development and deployment.

In conclusion, the integration of machine learning and robotics signifies a transformative juncture in the evolution of intelligent machines. The marriage of learning algorithms with robotic systems brings forth a new generation of machines that not only execute tasks with precision but also learn, adapt, and collaborate with humans. This symbiotic relationship between machine learning and robotics holds the promise of revolutionizing industries, healthcare, education, and research. As intelligent robots become increasingly integrated into our daily lives, the responsible development and ethical deployment of these technologies become imperative to ensure a future where machines and humans coexist harmoniously, leveraging the strengths of each for the betterment of society.

Cognitive Computing and Robotics

In the intricate dance of technological innovation, the convergence of cognitive computing and robotics stands as a milestone, propelling the capabilities of machines into realms that were once the domain of human intellect. The integration of cognitive computing, a branch of artificial intelligence (AI) focused on mimicking human thought processes, with robotics marks a transformative synergy that empowers machines not just to execute tasks but to understand, reason, and learn. This symbiotic relationship heralds a new era where robots become cognitive entities, ushering in a paradigm shift in the capabilities of automated systems.

At the heart of cognitive computing is the aspiration to replicate and augment human cognition. By leveraging machine learning, natural language processing, and advanced algorithms, cognitive computing systems acquire the ability to comprehend, interpret, and respond to information in a manner that mimics human intelligence. When applied to robotics, cognitive computing elevates machines from being mere mechanical executors of commands to intelligent entities capable of perceiving, reasoning, and adapting to dynamic environments.

Perception, a cornerstone of cognitive abilities, is revolutionized by the integration of cognitive computing in robotics. Computer vision algorithms, infused with cognitive capabilities, enable robots to not only "see" the world through cameras but also to understand and interpret visual information. This cognitive perception extends beyond mere object recognition; it involves context-aware understanding, allowing robots to discern the significance of objects in their surroundings. This heightened level of perception finds applications in diverse fields, from autonomous vehicles navigating complex urban landscapes to industrial robots identifying and interacting with objects in unstructured environments.

The marriage of cognitive computing and robotic perception extends to other senses, bringing a holistic understanding of the environment. Tactile sensors, equipped with cognitive capabilities, allow robots to feel and interact with objects, detecting textures, pressures, and temperatures. This cognitive sense of touch enhances the dexterity of robotic manipulation, enabling delicate interactions and precise object handling. The integration of cognitive perception allows robots to adapt their actions based on the tactile feedback received, creating a more nuanced and adaptive interaction with the physical world.

Decision-making, a quintessential aspect of human cognition, is transformed by cognitive computing in the realm of robotics. Planning and control algorithms, infused with cognitive capabilities, endow robots with the ability to optimize their movements, strategize tasks, and make decisions in complex and dynamic environments. Cognitive computing enables robots to reason about their actions, considering not only immediate objectives but also long-term goals and potential consequences. This cognitive decision-making is particularly valuable in scenarios where robots need to navigate uncertain or evolving conditions.

Machine learning, a subset of cognitive computing, introduces the concept of learning from experience into the realm of robotics. Reinforcement learning, a type of machine learning, allows robots to learn optimal behavior through trial and error. Robots equipped with cognitive learning algorithms can adapt their actions based on feedback from the environment, continuously improving their performance over time. This iterative learning process is particularly advantageous in scenarios where tasks are complex, and the optimal strategy may not be explicitly known.

Cognitive computing extends its influence into human-robot interaction, facilitating natural and intuitive communication. Natural language processing algorithms, a key component of cognitive computing, enable robots to understand and respond to human language. This capability goes beyond simple command recognition; it involves understanding context, intent, and even emotions conveyed through language. Robots equipped with cognitive language processing can engage in sophisticated dialogues with human users, making them more accessible and user-friendly. This cognitive interaction is crucial in applications where robots collaborate with humans, such as in customer service, healthcare, or education.

Collaborative robotics, emphasizing the coexistence of humans and robots in shared workspaces, benefits significantly from cognitive computing. Cognitive algorithms enable robots to recognize and respond to human behaviors, ensuring safe and efficient collaboration. These collaborative robots, often referred to as cobots, can adapt their movements and behaviors based on the actions of human coworkers, creating a harmonious and productive working environment. The cognitive understanding of human actions fosters a level of collaboration where robots not only execute tasks but actively engage with human collaborators.

Autonomy, a defining characteristic of cognitive computing, redefines the concept of robotic independence. Traditionally, robots were programmed with explicit instructions to perform specific tasks within predefined environments. Cognitive autonomy allows robots to learn, adapt, and make decisions in real-time, even in unpredictable and dynamic conditions. This level of autonomy is particularly crucial in applications where robots need to operate alongside humans, such as in collaborative manufacturing or service industries.

Swarm robotics, a collaborative approach inspired by the collective behavior observed in nature, leverages cognitive computing for decentralized decision-making. Swarms of small robots, communicating with each other through cognitive algorithms, can achieve complex tasks collectively. This decentralized intelligence allows swarms of robots to exhibit emergent behaviors and adapt to changing conditions. Cognitive swarm robotics finds applications in scenarios such as environmental monitoring, where a large number of distributed robots can cover vast areas efficiently, or in disaster response, where swarms of robots collaborate in search and rescue missions.

In healthcare, the integration of cognitive computing and robotics holds immense promise. Surgical robots, guided by cognitive algorithms, enhance the precision and capabilities of surgeons in minimally invasive procedures. Cognitive diagnostic robots analyze medical data, assist in the identification of diseases, and contribute to personalized treatment plans. Social robots, designed to interact with patients in healthcare settings, leverage cognitive computing for understanding and responding to human emotions, offering companionship and support. The cognitive capabilities of these robots contribute to improved medical procedures, diagnostics, and patient care.

The transformative impact of cognitive computing and robotics extends into education and research. Educational robotics kits, based on cognitive principles, provide students with hands-on experiences in building and programming robots. Researchers benefit from the adaptability and learning capabilities of cognitive algorithms, enabling them to prototype and test various robotic configurations without the need for extensive reprogramming.

Challenges accompany the integration of cognitive computing and robotics, spanning technical complexities to ethical considerations. The robustness and reliability of cognitive algorithms in real-world environments pose ongoing challenges, especially in scenarios with uncertainties and variations. Ensuring the explainability and transparency of AI-driven cognitive decisions becomes crucial, particularly in sensitive domains such as healthcare or criminal justice. Ethical considerations, including bias in cognitive algorithms and the impact on employment, necessitate careful scrutiny and the establishment of guidelines for responsible development and deployment.

In conclusion, the integration of cognitive computing and robotics signifies a revolutionary stride in the evolution of intelligent machines. The marriage of cognitive capabilities with robotic systems brings forth a new generation of machines that not only execute tasks with precision but also understand, reason, and learn from their experiences. This symbiotic relationship between cognitive computing and robotics holds the promise of transforming industries, healthcare, education, and research. As cognitive robots become increasingly integrated into our daily lives, the responsible development and ethical deployment of these technologies become imperative to ensure a future where machines and humans coexist harmoniously, leveraging the strengths of each for the betterment of society.

Challenges and Opportunities in AI-Driven Robotics

The fusion of artificial intelligence (AI) and robotics has given rise to a transformative wave that is reshaping industries, healthcare, and our daily lives. AI-driven robotics holds the promise of machines that not only execute predefined tasks but also learn, adapt, and interact with the world in increasingly sophisticated ways. However, this integration brings forth a host of challenges and opportunities that demand careful consideration as we navigate the complex landscape of intelligent automation.

One of the primary challenges in the realm of AI-driven robotics lies in the robustness and adaptability of AI algorithms in real-world environments. While AI models are often trained in controlled settings, the unpredictable nature of the physical world introduces complexities that can challenge the reliability of these algorithms. Variations in lighting conditions, unexpected obstacles, or changes in the environment pose hurdles that AI-driven robots must overcome. Ensuring that AI algorithms can adapt to these dynamic conditions and continue to operate safely and efficiently is an ongoing technical challenge.

The explainability of AI decisions is another critical consideration. As AI-driven robots make decisions based on complex algorithms and vast datasets, understanding the rationale behind these decisions becomes challenging. In sensitive domains such as healthcare or criminal justice, where AI-driven systems may influence critical outcomes, the ability to explain and interpret the decisions of intelligent machines becomes imperative. Striking a balance between the complexity of AI models and the need for transparency is an ongoing challenge in the responsible deployment of AI-driven robotics.

Ethical considerations loom large on the horizon of AI-driven robotics. Bias in AI algorithms, inadvertently learned from historical data, can perpetuate and even exacerbate societal inequalities. Ensuring fairness and mitigating bias in AI-driven systems requires vigilant efforts in data curation, algorithmic design, and ongoing monitoring. Additionally, the societal impact of automation on employment is a topic of ongoing concern. While AI-driven robotics can enhance productivity and efficiency, questions about job displacement and the redefinition of employment opportunities demand careful consideration and strategic planning.

Interdisciplinary collaboration emerges as both a challenge and an opportunity in the field of AI-driven robotics. The convergence of AI, robotics, and other cutting-edge technologies requires experts from diverse domains to work collaboratively. Engineers, computer scientists, ethicists, policymakers, and industry professionals must join forces to address the multifaceted challenges posed by AI-driven robotics. Interdisciplinary collaboration not only enhances the development of robust and responsible technologies but also fosters a holistic understanding of the implications and potential applications of AI-driven robotics.

Another challenge is the standardization of interfaces and communication protocols in AI-driven robotic systems. As the field advances, ensuring interoperability and compatibility between different robotic platforms and components becomes crucial. Standardization facilitates the integration of diverse robotic systems, promotes collaboration, and reduces barriers to entry for developers and researchers. Establishing common standards for AI-driven robotics is an ongoing effort that seeks to streamline the development and deployment of intelligent machines.

Privacy and security concerns emerge prominently in the deployment of AI-driven robotics, particularly in applications involving personal or sensitive data. As robots equipped with AI algorithms interact with the physical world and process information, ensuring the privacy of individuals becomes a paramount consideration. Implementing robust security measures to protect against unauthorized access and data breaches is essential to build trust in AI-driven robotic systems. Striking a balance between the benefits of data-driven intelligence and the protection of individual privacy is a nuanced challenge.

Despite these challenges, the opportunities presented by AI-driven robotics are vast and transformative. The enhanced capabilities of intelligent machines open new frontiers in industries such as manufacturing, healthcare, and logistics. AI-driven robots, equipped with advanced perception and decision-making capabilities, can optimize processes, improve efficiency, and contribute to innovation in diverse sectors. The ability of AI-driven robotics to handle complex tasks in unstructured environments positions them as valuable tools in applications ranging from disaster response to environmental monitoring.

The field of healthcare stands to benefit significantly from AI-driven robotics. Surgical robots, guided by AI algorithms, can perform intricate procedures with enhanced precision, reducing invasiveness and improving patient outcomes. AI-driven diagnostic robots contribute to the identification of diseases, analyze medical data, and support personalized treatment plans. Social robots, equipped with AI for understanding human emotions and interactions, find applications in patient care, rehabilitation, and companionship.

Education and research witness a transformative impact from AI-driven robotics. Educational robotics kits, based on AI principles, provide students with hands-on experiences in building and programming intelligent machines. Researchers leverage the adaptability and learning capabilities of AI algorithms to prototype and test various robotic configurations without extensive reprogramming. The integration of AI-driven robotics into academic and research settings accelerates the exploration of intelligent systems and fosters innovation. Collaborative robotics, exemplified by the coexistence of humans and robots in shared workspaces, presents an opportunity to redefine the nature of work. Cobots, equipped with AI for adaptive behaviors and safety features, enable humans and robots to collaborate seamlessly. This collaborative approach not only enhances efficiency in industrial settings but also contributes to a more flexible and responsive work environment. AI-driven robotics opens avenues for reimagining how humans and machines work together, leveraging the strengths of each to create a harmonious and productive partnership.

The democratization of AI-driven robotics is a notable opportunity that empowers developers, researchers, and businesses to access and leverage intelligent automation. Open-source frameworks, accessible development platforms, and shared resources contribute to the widespread adoption and advancement of AI- driven robotic technologies. The collaborative nature of the AI and robotics communities fosters innovation, accelerates research, and expands the possibilities of intelligent automation across diverse applications.

In conclusion, the journey through the challenges and opportunities in AI-driven robotics is a dynamic exploration of the frontiers of technology. While the integration of AI and robotics presents formidable challenges in terms of robustness, explainability, ethics, and interdisciplinary collaboration, the transformative potential is undeniable. The opportunities to revolutionize industries, redefine healthcare, enhance education, and reshape the nature of work underscore the profound impact of AI-driven robotics on society. As we navigate this complex landscape, it is imperative to approach the development and deployment of AI-driven robotic systems with a commitment to responsible innovation, ethical considerations, and a collaborative spirit that embraces the vast potential of intelligent automation.

CHAPTER IV

Robotic Applications in Industry

Industrial Automation

The landscape of modern industry is undergoing a seismic transformation, driven by the relentless march of technological progress. At the forefront of this revolution stands industrial automation, a multidimensional paradigm that integrates cutting-edge technologies to streamline processes, enhance efficiency, and redefine the very nature of production. The convergence of robotics, artificial intelligence (AI), and advanced sensor technologies has given rise to a new era where factories and manufacturing facilities are no longer bound by the limitations of traditional manual labor but are instead propelled by the precision and adaptability of intelligent machines.

A cornerstone of industrial automation is the integration of robotics into manufacturing processes. Robots, equipped with advanced sensors and programmed with sophisticated algorithms, have become integral to tasks that demand precision, speed, and repeatability. From assembly lines in automotive manufacturing to intricate tasks in electronics production, robots are adept at handling a wide array of responsibilities. The versatility of robotic systems extends beyond routine tasks, as they increasingly collaborate with human workers in shared workspaces, demonstrating adaptability and intelligence in dynamic environments.

Artificial intelligence serves as the cognitive powerhouse behind industrial automation, imbuing machines with the ability to learn, reason, and make decisions. Machine learning algorithms analyze vast datasets to optimize processes, predict maintenance needs, and improve overall efficiency. In predictive maintenance, for example, AI algorithms can anticipate equipment failures before they occur, enabling proactive interventions that minimize downtime and enhance the reliability of industrial systems. The marriage of AI and industrial automation not only elevates the capabilities of machines but also introduces a level of adaptability that is crucial in the face of dynamic production environments.

Sensors play a pivotal role in the sensory fabric of industrial automation, providing machines with the ability to perceive and respond to their surroundings. Advanced sensors, ranging from vision systems and proximity sensors to temperature and pressure sensors, enable machines to gather real-time data. This sensory input allows machines to adapt their actions based on environmental conditions, ensuring precision and reliability in manufacturing processes. The integration of sensors in industrial automation fosters a data-driven approach, where decision-making is informed by a continuous stream of information, leading to optimized performance and resource utilization.

The impact of industrial automation is particularly evident in the realm of smart factories, where interconnected systems and the Internet of Things (IoT) create a networked ecosystem of intelligent devices. In a smart factory, machines communicate with each other, share data, and collaboratively optimize production processes. IoT-enabled sensors monitor equipment performance, track inventory levels, and provide real- time insights into production metrics. This interconnectedness facilitates a holistic view of the manufacturing ecosystem, enabling rapid decision- making and adaptive responses to changing conditions.

The benefits of industrial automation extend far beyond the confines of the factory floor, permeating supply chain management and logistics. Automated systems, guided by AI algorithms, optimize inventory levels, forecast demand, and streamline the movement of goods. Autonomous vehicles and drones, equipped with automation technologies, navigate warehouses and distribution centers with precision, reducing the need for manual intervention. The integration of automation in logistics not only improves efficiency but also contributes to sustainability efforts by minimizing energy consumption and optimizing transportation routes. One of the fundamental advantages of industrial automation lies in its capacity to enhance overall efficiency and productivity. Machines, driven by automation technologies, can operate continuously without the need for breaks, resulting in increased output and reduced cycle times. This heightened efficiency is particularly critical in industries where precision and speed are paramount, such as semiconductor manufacturing or pharmaceutical production. The ability of automated systems to handle repetitive tasks with consistency and accuracy leads to a reduction in errors and an improvement in the overall quality of manufactured products.

Labor-intensive tasks, often associated with manual assembly and production, are gradually being delegated to automated systems. This transition allows human workers to focus on more complex and intellectually demanding aspects of production, such as system design, programming, and maintenance. The symbiotic collaboration between humans and machines in industrial settings exemplifies the potential for automation to augment human capabilities rather than replace them. This paradigm shift in the role of human workers in industrial processes contributes to a more dynamic and fulfilling work environment.

Safety considerations are paramount in industrial automation, with advancements in technology contributing to the development of safer workspaces. Collaborative robots, often referred to as cobots, are designed to work alongside human operators in shared environments. Equipped with sensors and safety features, cobots can adapt their movements and behaviors in response to human presence, minimizing the risk of accidents. Automation technologies also contribute to hazardous task mitigation, as machines can be deployed in environments where conditions may pose a danger to human workers.

The deployment of industrial automation has a profound impact on the economic competitiveness of nations and industries. Increased efficiency, reduced production costs, and enhanced product quality contribute to the overall competitiveness of manufacturing sectors. Nations that embrace and invest in industrial automation technologies position themselves at the forefront of global economic competition. The strategic adoption of automation not only boosts productivity but also fosters innovation, laying the foundation for sustained economic growth.

While the benefits of industrial automation are vast, challenges persist in its widespread adoption. The initial costs associated with implementing automation technologies, including the acquisition of robotic systems and the integration of AI, can be substantial. Small and medium-sized enterprises (SMEs), in particular, may face barriers to entry due to financial constraints and the need for specialized expertise. Overcoming these challenges requires strategic planning, government incentives, and the development of accessible automation solutions tailored to the needs of diverse industries.

Workforce reskilling emerges as a critical consideration in the era of industrial automation. As the nature of jobs evolves with the integration of intelligent machines, there is a growing demand for workers with skills in robotics programming, system maintenance, and data analytics. Ensuring that the workforce is equipped with the necessary skills to navigate the automated landscape is imperative for a smooth transition. Education and training programs, both within academic institutions and through industry partnerships, play a pivotal role in preparing the workforce for the challenges and opportunities presented by industrial automation. Ethical considerations accompany the rise of industrial automation, particularly in the context of job displacement and societal impact. The automation of routine tasks may lead to shifts in employment patterns, raising questions about the role of workers in the automated economy. Addressing these ethical considerations requires a nuanced approach that prioritizes the well-being of workers, fosters inclusive economic growth, and acknowledges the broader societal implications of industrial automation.
In conclusion, industrial automation represents a transformative force that is reshaping the foundations of manufacturing and production. The integration of

robotics, artificial intelligence, and advanced sensors propels industries into a new era of efficiency, adaptability, and innovation. While challenges such as initial costs, workforce reskilling, and ethical considerations persist, the opportunities presented by industrial automation are too compelling to ignore. As we navigate this era of intelligent machines, the responsible deployment of automation technologies, coupled with strategic investments in education and training, will be instrumental in realizing the full potential of industrial automation and driving sustained economic growth.

Robotics in Manufacturing

The integration of robotics into manufacturing processes stands as a hallmark of technological progress, ushering in a new era where precision, efficiency, and adaptability converge to redefine the landscape of production. Across industries, from automotive assembly lines to electronics manufacturing, robots have become indispensable collaborators in the quest for heightened productivity and quality. The synergy of robotics and manufacturing technologies has not only streamlined traditional production paradigms but has also opened new frontiers in innovation, laying the foundation for the factories of the future.

Central to the transformative impact of robotics in manufacturing is the versatility and precision with which robots can execute tasks. Traditionally relegated to repetitive and mundane activities, robots now navigate intricate assembly processes with dexterity and speed. In automotive manufacturing, for instance, robotic arms seamlessly weld, paint, and assemble components, achieving levels of precision unattainable through manual labor. The adaptability of robotic systems allows for rapid reconfiguration, enabling manufacturers to swiftly switch between production setups and respond to changing demands in the market.

The advent of collaborative robots, or cobots, exemplifies the evolution of human-robot collaboration on the manufacturing floor. Cobots are designed to work alongside human operators, leveraging advanced sensors and safety features to ensure a harmonious and safe coexistence. In shared workspaces, cobots execute tasks that demand a delicate touch or intricate manipulation, freeing human workers to focus on tasks that require creativity, problem-solving, and complex decision-making. This collaborative approach not only enhances efficiency but also contributes to a more dynamic and engaging work environment.

Artificial intelligence (AI) plays a pivotal role in enhancing the capabilities of robots in manufacturing. Machine learning algorithms empower robots to learn from data, adapt to variations in production processes, and optimize their performance over time. In predictive maintenance, AI-driven systems analyze data from sensors to forecast equipment failures, enabling proactive interventions that minimize downtime and extend the lifespan of machinery. The fusion of AI and robotics creates intelligent manufacturing systems that continuously learn and evolve, contributing to a paradigm shift in how factories operate.

Vision systems, another integral component of robotic technology, elevate the perceptual abilities of machines in manufacturing environments. Equipped with cameras and image recognition algorithms, robots can "see" and interpret their surroundings with a level of sophistication that rivals human perception. In quality control processes, vision systems inspect products for defects with unmatched precision, ensuring that only products meeting stringent quality standards reach the market. The integration of vision systems into robotic workflows contributes to the reliability and consistency of manufacturing processes.

The impact of robotics in manufacturing extends beyond routine assembly tasks to intricate operations such as material handling and logistics. Automated guided vehicles (AGVs) and autonomous mobile robots (AMRs) navigate warehouse floors, transporting materials and components with efficiency and accuracy. The deployment of robotic systems in logistics optimizes supply chain management, reduces lead times, and minimizes errors in inventory management. As a result, manufacturers can achieve a level of operational efficiency that was once inconceivable, contributing to the overall competitiveness of their enterprises.

Sensors, ranging from proximity sensors to force sensors, play a pivotal role in enhancing the interaction between robots and their environments. Force sensors, for example, enable robots to apply controlled pressure and adapt their movements based on the resistance encountered during tasks such as assembly or material handling. Tactile sensors, mimicking the sense of touch, allow robots to interact delicately with objects and respond to variations in surface textures. The integration of these advanced sensors contributes to the refinement of robotic movements, expanding the range of tasks that machines can undertake with precision.

The concept of lights-out manufacturing, where factories operate autonomously without human intervention, exemplifies the transformative potential of robotics in manufacturing. In lights-out facilities, robotic systems execute tasks 24/7, achieving unprecedented levels of efficiency and productivity. While lights-out manufacturing is not universally applicable, its emergence in certain industries underscores the extent to which robotic technologies can redefine traditional notions of production cycles and operational timelines.

The deployment of robotics in manufacturing contributes to advancements in the customization and personalization of products. Flexible robotic systems, capable of rapid reprogramming and adaptation, enable manufacturers to respond swiftly to changing consumer preferences and market demands. In industries such as electronics, where product life cycles are short, the agility of robotic systems ensures that production processes can be quickly reconfigured to introduce new features or accommodate design changes. This flexibility positions manufacturers to navigate the dynamic landscape of consumer preferences with ease.

The impact of robotics in manufacturing is not confined to large-scale enterprises; small and medium-sized enterprises (SMEs) also stand to benefit. Collaborative robots, designed for ease of use and rapid deployment, offer SMEs the opportunity to integrate automation into their production processes without substantial investments in infrastructure or specialized expertise. The accessibility of robotic technologies, coupled with the availability of open-source platforms and collaborative ecosystems, fosters a democratization of automation, enabling businesses of varying scales to leverage the advantages of robotic systems.

Despite the transformative potential of robotics in manufacturing, challenges persist in the widespread adoption of these technologies. Initial costs associated with the acquisition and integration of robotic systems can be a barrier for some manufacturers, particularly SMEs with limited capital. Overcoming these cost challenges requires strategic planning, government incentives, and the development of affordable automation solutions tailored to the needs of diverse industries.

The redefinition of the role of human workers in the era of robotic manufacturing is a multifaceted consideration. As robots take on routine and physically demanding tasks, human workers are liberated to focus on activities that require creativity, problem-solving, and adaptability. This shift in the division of labor underscores the potential for human-robot collaboration to create more fulfilling and dynamic work environments. However, workforce reskilling emerges as a critical imperative to ensure that workers possess the skills necessary to navigate the evolving landscape of automated manufacturing.

Ethical considerations accompany the integration of robotics in manufacturing, particularly in the context of job displacement and societal impact. Ensuring a just transition for workers affected by automation, fostering inclusive economic growth, and addressing concerns related to job quality and security demand careful attention. Ethical deployment of robotics requires a commitment to responsible innovation, acknowledging the broader societal implications of automation and actively engaging with stakeholders to mitigate potential negative consequences.

In conclusion, robotics in manufacturing represents a paradigm shift that transcends traditional production methods, ushering in an era of unprecedented efficiency, precision, and adaptability. The collaborative dance of robots and humans on the factory floor exemplifies the potential for technology to enhance the capabilities of both, creating a synergy that drives innovation and competitiveness. As robotics continues to evolve, the responsible deployment of these technologies, coupled with strategic investments in education, training, and ethical considerations, will be instrumental in maximizing the benefits of robotic manufacturing and ensuring a future where machines and humans coexist harmoniously in the pursuit of progress.

Logistics and Warehousing

In the dynamic landscape of modern industry, the marriage of robotics and logistics has ushered in a new era of efficiency, precision, and adaptability in supply chain management. The convergence of robotic applications with logistics and warehousing processes is reshaping the way goods are handled, stored, and transported, optimizing every facet of the supply chain. From automated guided vehicles (AGVs) to robotic arms sorting packages, the integration of robotics in logistics and warehousing operations is revolutionizing the industry, propelling it into a future where speed, accuracy, and flexibility are paramount.

A central player in the realm of robotic applications in logistics is the deployment of AGVs and autonomous mobile robots (AMRs). These intelligent robotic vehicles navigate warehouse floors with precision, transporting goods from one location to another in a seamless and efficient manner. In large-scale distribution centers, AGVs follow predefined paths or navigate autonomously using sensors and mapping technologies, reducing the need for manual labor in material handling. The ability of AGVs to operate continuously and adapt to changing warehouse layouts contributes to streamlined internal logistics, minimizing bottlenecks and optimizing the flow of goods.

Robotics in warehousing extends beyond basic material handling to intricate processes such as order picking and packing. Robotic arms equipped with advanced vision systems and gripping mechanisms are increasingly employed to pick items from shelves and pack them into boxes. The integration of vision systems allows these robotic arms to identify and grasp objects with a level of precision that rivals human dexterity. This application of robotics in order fulfillment processes not only accelerates the speed of operations but also minimizes errors, enhancing the overall accuracy of order processing.

Automated storage and retrieval systems (AS/RS) represent a pinnacle in the fusion of robotics and warehousing. These systems, often employed in large-scale distribution centers, utilize robotic cranes and conveyors to autonomously retrieve items from densely packed storage units. The integration of AS/RS eliminates the need for manual searching and handling of goods in warehouses with high-density storage, significantly reducing retrieval times and optimizing storage space utilization. The precision and speed of AS/RS contribute to efficient order fulfillment, particularly in industries characterized by a high volume of SKUs and rapid turnover.

In e-commerce fulfillment centers, where the demand for rapid order processing is paramount, robotics play a transformative role in achieving operational excellence. Goods-to-person robotic systems bring items directly to human operators, minimizing the time spent walking through aisles and maximizing the efficiency of order picking. This collaborative approach, where robots and humans work in tandem, exemplifies the potential for robotics to enhance the capabilities of human workers and streamline complex logistics operations. The result is a faster and more responsive supply chain, meeting the expectations of consumers for swift and accurate deliveries.

The integration of robotics in logistics is particularly evident in the last-mile delivery, the final leg of the supply chain that involves delivering goods to the end consumer. Autonomous delivery vehicles, including drones and ground-based robots, navigate urban landscapes to fulfill last-mile deliveries. Drones, equipped with sensors and GPS technology, can traverse challenging terrains to reach remote locations, providing a solution for quick and efficient deliveries in diverse environments. Ground-based robots, designed for sidewalk or road navigation, contribute to contactless and timely delivery services, especially in densely populated urban areas.

The advent of the Internet of Things (IoT) further amplifies the impact of robotics in logistics and warehousing. IoT-connected sensors and devices create a networked ecosystem where real-time data is collected and shared across the supply chain. This connectivity enhances visibility and transparency, allowing stakeholders to monitor the status and location of goods throughout the entire logistics process. The integration of robotics with IoT technologies facilitates data-driven decision-making, enabling proactive responses to disruptions, optimizing inventory levels, and enhancing overall supply chain resilience.

Warehousing and logistics robotics are not confined to traditional indoor environments. Automated yard management systems leverage robotic technologies to optimize the movement and storage of goods in outdoor areas, such as distribution center yards and ports. Autonomous vehicles, guided by robotics and IoT technologies, navigate these outdoor spaces, coordinating the transfer of containers and streamlining yard operations. The application of robotics in yard management contributes to overall supply chain efficiency by minimizing congestion, reducing wait times, and improving the utilization of outdoor storage areas.

The impact of robotics in logistics is not limited to the optimization of operational processes; it extends to sustainability initiatives within the industry. Electric-powered AGVs and robotic vehicles contribute to reduced emissions and lower energy consumption compared to traditional fuel-powered equipment. The optimization of routes and the reduction of idle times achieved through robotics in logistics also contribute to a greener supply chain. As sustainability becomes a key focus for industries worldwide, the integration of robotics aligns with environmental goals, creating a more efficient and eco-friendly logistics landscape. Challenges accompany the integration of robotics in logistics and warehousing, spanning technical complexities to workforce considerations. The initial investment in robotics technologies, including the acquisition of robotic systems and the implementation of necessary infrastructure, can be a substantial barrier for some logistics providers. Overcoming these challenges requires strategic planning, cost-benefit analysis, and the development of flexible solutions that cater to the specific needs of diverse logistics operations. Workforce considerations in the era of logistics robotics encompass a nuanced balance between human workers and intelligent machines. While robotics automate routine and physically demanding tasks, human workers remain integral to decision-making, complex problem-solving, and tasks that require creativity. The collaborative approach, where robots and humans work together in shared spaces, demands a shift in the skill set of the workforce. Training programs and reskilling initiatives become imperative to equip workers with the necessary expertise to operate, monitor, and maintain robotic systems effectively.

Ethical considerations in logistics robotics revolve around job displacement and the broader societal impact of automation. Addressing these concerns requires a comprehensive approach that includes social dialogue, proactive workforce planning, and the development of policies that prioritize worker well-being. Ethical deployment of robotics in logistics involves a commitment to responsible innovation, ensuring that the benefits of automation are shared equitably among workers and that societal impacts are actively managed. In conclusion, the integration of robotics in logistics and warehousing is reshaping the foundations of supply chain management, creating a landscape where efficiency, precision, and adaptability are paramount. From the optimized movement of goods within warehouses to the streamlined last-mile deliveries, robotics contribute to a more responsive and agile supply chain. As industries embrace the transformative potential of logistics robotics, it becomes imperative to navigate the associated challenges with strategic planning, workforce development, and ethical considerations. In doing so, the synergy of robotics and logistics holds the promise of not only enhancing operational efficiency but also creating a supply chain ecosystem that is sustainable, resilient, and responsive to the demands of the modern world.

Economic Impact and Efficiency Gains

The integration of robotic applications into industrial processes has unleashed a wave of economic transformation, fundamentally altering the landscape of manufacturing, logistics, and beyond. As intelligent machines become integral to diverse sectors, from automotive production lines to e-commerce fulfillment centers, the economic impact and efficiency gains are reshaping traditional paradigms and driving industries towards new frontiers of productivity.

At the heart of the economic impact of robotics lies the pursuit of efficiency gains. Robots, equipped with advanced sensors, artificial intelligence (AI), and robotic vision systems, are capable of executing tasks with a level of precision and speed that surpasses human capabilities. In manufacturing, this translates to streamlined production processes, reduced cycle times, and enhanced overall efficiency. Robotic arms on assembly lines, guided by AI algorithms, perform intricate tasks with consistency, contributing to a reduction in errors and an improvement in the quality of manufactured products.

The efficiency gains realized through robotic applications extend beyond the factory floor to logistics and supply chain management. Automated guided vehicles (AGVs) navigate warehouse spaces with precision, optimizing material handling processes and reducing the reliance on manual labor. Robotics in warehousing, including robotic arms for order picking and autonomous mobile robots (AMRs) for internal logistics, contribute to accelerated order fulfillment and minimized errors in distribution centers. The net result is a supply chain ecosystem characterized by responsiveness, reduced lead times, and enhanced overall operational efficiency.

In the realm of economic impact, the adoption of robotics is a catalyst for competitiveness on a global scale. Nations and industries that strategically invest in robotic technologies position themselves at the forefront of innovation and efficiency. The ability of robotic systems to operate continuously, adapt to changing production requirements, and optimize resource utilization contributes to the overall competitiveness of manufacturing sectors. The economic benefits extend to enhanced product quality, reduced production costs, and the potential for increased market share in a globally connected economy.

The economic impact of robotics is particularly pronounced in the context of Industry 4.0, the fourth industrial revolution characterized by the integration of digital technologies, automation, and data exchange in manufacturing. Smart factories, where interconnected systems and intelligent machines collaborate seamlessly, exemplify the transformative potential of Industry 4.0. Robotics serves as a cornerstone in the realization of smart manufacturing, contributing to the creation of flexible, agile, and data-driven production environments. The economic implications of Industry 4.0 extend beyond individual enterprises to national economies, fostering innovation, driving productivity, and contributing to sustained economic growth.

Labor productivity, a key economic metric, experiences a paradigm shift with the integration of robotics. While concerns about job displacement are valid, the augmentation of human capabilities through the collaboration with robots results in a more dynamic and efficient workforce. Routine and physically demanding tasks are delegated to robotic systems, freeing human workers to focus on activities that require creativity, critical thinking, and complex decision-making. The synergy of human and machine labor not only enhances overall productivity but also contributes to a more fulfilling and engaging work environment.

The economic impact of robotics in manufacturing is evident in the concept of lights-out manufacturing, where factories operate autonomously without human intervention. In lights-out facilities, robotic systems execute tasks 24/7, achieving unprecedented levels of efficiency and productivity. While lights-out manufacturing is not universally applicable, its emergence in certain industries underscores the extent to which robotic technologies can redefine traditional notions of production cycles and operational timelines. The economic benefits include minimized labor costs, optimized energy consumption, and the potential for continuous operation in high-demand industries.

The efficiency gains facilitated by robotics in logistics and warehousing contribute to substantial economic benefits. E-commerce fulfillment centers, faced with the challenge of meeting consumer expectations for swift and accurate deliveries, leverage robotic technologies to optimize order processing. Goods-to-person robotic systems, AGVs, and robotic arms work collaboratively to expedite order fulfillment, reducing operational costs and enhancing customer satisfaction. The economic impact extends to reduced shipping costs, minimized storage expenses, and the ability to navigate the complexities of modern supply chain demands.
In the context of small and medium-sized enterprises (SMEs), the economic impact of robotics is a transformative force that levels the playing field.

Collaborative robots, or cobots, designed for ease of use and rapid deployment, offer SMEs the opportunity to integrate automation into their production processes without substantial investments in infrastructure or specialized expertise. The accessibility of robotic technologies, coupled with the availability of open- source platforms and collaborative ecosystems, fosters a democratization of automation. SMEs can leverage the economic benefits of robotics to enhance competitiveness, drive innovation, and achieve operational excellence.

The economic impact of robotics extends to job creation, albeit in roles that require new skills and expertise. As industries embrace automation, the demand for workers with skills in robotics programming, system maintenance, and data analytics rises. Educational institutions and training programs play a pivotal role in equipping the workforce with the necessary skills to navigate the evolving landscape of automated industries. The economic benefits of job creation align with the broader goal of fostering a skilled and adaptive workforce that can contribute to the sustained growth of industries embracing robotic technologies.

However, the economic impact of robotics also poses challenges that demand careful consideration. The initial costs associated with the acquisition and integration of robotic systems can be a barrier for some enterprises, particularly SMEs with limited capital. Overcoming these cost challenges requires strategic planning, government incentives, and the development of affordable automation solutions tailored to the needs of diverse industries. The return on investment (ROI) in robotics, while significant in the long term, requires a nuanced approach to justify initial investments and ensure economic sustainability.

Ethical considerations accompany the economic impact of robotics, particularly in the context of job displacement and societal impact. The responsible deployment of robotics involves proactive workforce planning, reskilling initiatives, and the development of policies that prioritize worker well-being. Ethical considerations extend to issues of equity, diversity, and inclusion in the workforce affected by automation. Striking a balance between the economic benefits of robotics and the ethical responsibility to mitigate negative consequences requires a collaborative and forward-thinking approach from industries, policymakers, and stakeholders.

In conclusion, the economic impact and efficiency gains realized through robotic applications in industry are reshaping the foundations of production, logistics, and manufacturing. From streamlined assembly lines to responsive supply chains, robotics contributes to a future where productivity, innovation, and competitiveness are synonymous with intelligent automation. As industries navigate the transformative potential of robotics, it becomes imperative to address challenges with strategic planning, workforce development, and ethical considerations. In doing so, the economic impact of robotics becomes a driving force for sustainable growth, innovation, and the evolution of industries towards a future defined by the seamless collaboration of humans and machines.

CHAPTER V

Robotics in Healthcare

Surgical Robotics

The field of surgery has undergone a profound transformation with the advent of surgical robotics, ushering in a new era of precision, minimally invasive procedures, and expanded capabilities for healthcare professionals. Surgical robots represent a convergence of cutting-edge technologies, including robotics, artificial intelligence (AI), and advanced imaging, to redefine the way surgeries are performed. From intricate procedures to routine interventions, surgical robotics have become indispensable tools in the hands of surgeons, contributing to improved patient outcomes, reduced recovery times, and enhanced overall surgical capabilities.

At the core of surgical robotics is the drive for precision in medical interventions. Traditional open surgeries, characterized by large incisions and direct manual manipulation, have given way to minimally invasive procedures facilitated by robotic systems. The da Vinci Surgical System, one of the pioneering platforms in surgical robotics, exemplifies this shift. Comprising robotic arms controlled by a console, the da Vinci system allows surgeons to perform complex surgeries through small incisions with unmatched precision. The articulation of robotic instruments, mimicking the range of human hand movements, enables surgeons to navigate anatomical structures with dexterity, reducing trauma to surrounding tissues and enhancing the overall precision of surgical procedures.

Minimally invasive surgeries facilitated by robotic systems contribute to reduced recovery times and improved patient outcomes. Smaller incisions result in less postoperative pain, reduced risk of infection, and shorter hospital stays. The enhanced precision afforded by robotic instruments enables surgeons to perform intricate procedures with greater accuracy, particularly in areas of the body where precision is critical, such as neurosurgery or delicate urological interventions. The application of surgical robotics in oncological surgeries, such as prostatectomies and gynecological procedures, has become standard practice, offering patients the benefits of both efficacy and reduced postoperative morbidity.

The integration of artificial intelligence into surgical robotics introduces a cognitive dimension to medical interventions. Machine learning algorithms, trained on vast datasets of surgical procedures, enhance the capabilities of robotic systems. In preoperative planning, AI algorithms analyze patient-specific data, such as imaging scans, to assist surgeons in developing personalized surgical approaches. During surgery, AI-driven assistance provides real-time insights, helping surgeons make informed decisions based on a continuous stream of data. The marriage of AI and surgical robotics not only augments the skills of surgeons but also contributes to a data-driven approach to medical interventions, fostering a new frontier in personalized and precise healthcare.

One of the defining features of surgical robotics is the teleoperated or remote-controlled nature of these systems. Surgeons manipulate robotic instruments from a console, often situated away from the operating table. This remote-control capability introduces a level of flexibility that transcends geographical constraints, enabling expert surgeons to perform interventions across vast distances. Telesurgery, facilitated by surgical robotics, has the potential to address healthcare disparities by bringing specialized surgical expertise to underserved regions. However, the realization of telesurgery also poses challenges, including issues of real-time communication, data security, and regulatory considerations that require careful navigation.

Surgical robotics has extended its reach beyond traditional procedures to encompass a diverse range of specialties. In orthopedic surgeries, robotic-assisted systems aid in tasks such as joint replacement procedures, ensuring precise alignment and optimal placement of implants. In cardiovascular interventions, robotic platforms contribute to the precision of delicate procedures such as coronary artery bypass surgeries. The versatility of surgical robotics extends to ophthalmic surgeries, where robots assist in delicate eye surgeries with micron-level precision. The adaptability of robotic systems across multiple specialties reflects the transformative potential of these technologies in reshaping the landscape of surgical interventions. In the realm of pediatric surgery, where the anatomical nuances of young patients present unique challenges, surgical robotics offer a paradigm shift. The precision and minimally invasive nature of robotic instruments contribute to reduced trauma in pediatric surgeries, minimizing the impact on developing tissues. The articulation of robotic arms, controlled by surgeons with heightened precision, allows for intricate procedures in pediatric urology, general surgery, and even cardiac interventions. The application of surgical robotics in pediatric settings underscores the adaptability of these technologies across diverse patient populations.

Despite the transformative potential of surgical robotics, challenges persist in their widespread adoption. The initial costs associated with acquiring and implementing robotic systems can be substantial, posing financial barriers for some healthcare institutions. The return on investment (ROI) in surgical robotics requires careful consideration of factors such as procedural volumes, patient outcomes, and operational efficiencies. Striking a balance between the potential benefits and the economic considerations is a critical aspect of the strategic integration of surgical robotics into healthcare settings.

Training and education emerge as pivotal components in the successful deployment of surgical robotics. Surgeons, nurses, and operating room staff require specialized training to master the nuances of robotic-assisted procedures. Simulation-based training programs, augmented reality platforms, and hands-on experiences are essential to ensure that healthcare professionals are proficient in leveraging the capabilities of surgical robotics. The integration of robotics into medical curricula becomes imperative to prepare the next generation of surgeons for the evolving landscape of surgical interventions.

Ethical considerations accompany the integration of surgical robotics, particularly in the context of patient safety, informed consent, and the role of human expertise. The reliance on robotic systems raises questions about the level of autonomy granted to machines in medical interventions. Ensuring transparency in the decision-making process, maintaining the primacy of the surgeon's expertise, and fostering open communication with patients become ethical imperatives in the era of surgical robotics. Striking a balance between technological advancements and ethical considerations is crucial to building trust in the use of robotic systems in healthcare.

In conclusion, surgical robotics stands as a transformative force in the field of medicine, reshaping the way surgeries are performed and opening new frontiers in precision healthcare. From the teleoperated control of robotic instruments to the integration of artificial intelligence, surgical robotics represents a convergence of technologies that holds the promise of improved patient outcomes and expanded capabilities for healthcare professionals. As surgical robotics continue to evolve, addressing challenges related to costs, training, and ethical considerations becomes imperative to ensure the responsible and widespread integration of these technologies. In the dynamic intersection of technology and medicine, surgical robotics emerges as a pioneering force, pushing the boundaries of what is possible in the pursuit of precision and excellence in surgical interventions.

Assistive Technologies

Assistive technologies represent a transformative force in the realm of healthcare and accessibility, revolutionizing the way individuals with disabilities navigate the world. Defined as tools, devices, or systems designed to enhance the independence and quality of life for people with disabilities, assistive technologies span a diverse range of applications, from mobility aids to communication devices. The overarching goal of these technologies is to break down barriers, empower individuals with disabilities, and foster inclusivity in society.

One of the most impactful categories of assistive technologies is mobility aids, catering to individuals with varying degrees of mobility impairments. Wheelchairs, both manual and powered, offer individuals with mobility challenges the freedom to move independently. Advances in technology have given rise to smart wheelchairs equipped with sensors and navigation systems, enabling users to control their movements with greater precision. Beyond traditional wheelchairs, exoskeletons represent a cutting-edge development, providing powered support to enhance mobility for individuals with paralysis or weakness. These wearable robotic devices enable users to stand, walk, and regain a level of independence that was once thought impossible. Communication is a fundamental aspect of human interaction, and assistive technologies have played a pivotal role in providing communication solutions for individuals with speech and language impairments. Augmentative and alternative communication (AAC) devices, ranging from simple picture boards to sophisticated speech-generating devices, empower individuals with conditions such as cerebral palsy or ALS to express themselves effectively. Eye-tracking technology has further expanded communication possibilities, allowing individuals to control devices and type messages using eye movements. These advancements in communication assistive technologies not only facilitate self-expression but also contribute to social inclusion and meaningful engagement with the world.

Visual impairments pose unique challenges, but assistive technologies have emerged as powerful tools to enhance accessibility for individuals with blindness or low vision. Screen readers, software that converts digital text into synthesized speech, enable users to navigate websites, documents, and digital interfaces. Braille displays provide tactile feedback, allowing individuals to read and interact with digital content using raised Braille dots. Navigation aids, such as smart canes equipped with sensors and GPS, enhance mobility and spatial awareness. The integration of artificial intelligence has further elevated the capabilities of assistive technologies for visual impairments, enabling real-time object recognition and scene description.

Hearing impairments find solutions in assistive technologies designed to enhance auditory experiences and communication. Hearing aids, worn in or behind the ear, amplify sound for individuals with hearing loss. Cochlear implants, a groundbreaking innovation, provide a direct interface with the auditory nerve for individuals with profound hearing loss. Communication devices equipped with text-to-speech capabilities facilitate seamless interaction for individuals with hearing impairments, ensuring effective communication in various settings. The evolution of assistive technologies for hearing impairments reflects a commitment to fostering inclusive communication and participation in diverse social contexts.

Cognitive and learning disabilities are addressed through assistive technologies that cater to individualized learning needs. Text-to-speech and speech-to-text applications provide alternatives for processing and expressing information, supporting individuals with dyslexia or other reading challenges. Educational software with customizable settings accommodates various learning styles and preferences. Cognitive aids, including reminder apps and organizational tools, assist individuals with attention and memory difficulties. The customization and flexibility offered by these assistive technologies empower individuals with cognitive disabilities to pursue education and professional opportunities with increased confidence.

Assistive technologies extend their impact to activities of daily living, fostering independence and autonomy for individuals with disabilities. Smart home systems, integrated with voice control and automation, enable individuals with mobility impairments to manage household tasks independently. Adaptive utensils and devices enhance eating and cooking experiences for individuals with dexterity challenges. Environmental control systems allow users to adjust lighting, temperature, and electronic devices with simple interfaces or voice commands. The convergence of assistive technologies with the Internet of Things (IoT) creates a connected ecosystem that adapts to individual needs, promoting self-sufficiency in daily life.

In the workplace, assistive technologies are instrumental in creating inclusive environments and unlocking the potential of individuals with disabilities. Screen magnifiers and screen readers enable individuals with visual impairments to access digital documents and participate in computer-based tasks. Ergonomic accommodations, including adjustable desks and specialized chairs, address the needs of individuals with physical disabilities. Speech recognition software facilitates hands-free computer interaction, benefiting individuals with mobility challenges. These workplace adaptations underscore the role of assistive technologies in breaking down barriers and creating opportunities for meaningful employment.

The advent of wearable technologies has expanded the horizons of assistive applications, providing discrete and integrated solutions for various disabilities. Wearable devices equipped with sensors and haptic feedback assist individuals with autism spectrum disorders in managing sensory challenges. Smart glasses with augmented reality capabilities enhance visual information for individuals with low vision. Wearable exoskeletons and robotic devices contribute to enhanced mobility for individuals with paralysis or muscle weakness. The discreet nature of wearables aligns with the principles of dignity and user preference, ensuring that assistive technologies seamlessly integrate into daily life.

Despite the transformative impact of assistive technologies, challenges persist in ensuring equitable access and widespread adoption. Affordability remains a significant barrier for many individuals with disabilities, particularly in regions with limited resources. Advocacy for policy changes, insurance coverage, and public awareness becomes imperative to address these financial barriers and ensure that assistive technologies are accessible to all who need them. Additionally, standardization and interoperability of assistive technologies are critical considerations to enhance compatibility, ease of use, and the seamless integration of different devices and systems.

The development of assistive technologies requires a user-centered approach, involving individuals with disabilities in the design and testing phases. User feedback and collaboration ensure that assistive technologies meet the diverse and evolving needs of the disability community. Furthermore, ongoing research and innovation are essential to push the boundaries of what is possible, fostering continuous improvement and the development of new solutions to address unmet needs.

In conclusion, assistive technologies stand as beacons of innovation, reshaping the lives of individuals with disabilities and fostering a more inclusive and accessible world. From mobility aids to communication devices, these technologies embody the spirit of empowerment, providing tools that enable individuals to overcome barriers and participate fully in society. The ongoing evolution of assistive technologies holds the promise of even greater advancements, driven by a commitment to innovation, user collaboration, and the vision of a world where abilities define potential, not limitations.

Rehabilitation Robotics

Rehabilitation robotics represents a groundbreaking frontier in the field of healthcare, where technology intersects with the human spirit's resilience to overcome physical challenges. Designed to assist individuals in regaining lost functionality, rehabilitation robots offer a spectrum of applications, from aiding in the recovery of motor skills to providing support for activities of daily living. As a transformative force, rehabilitation robotics not only accelerates the rehabilitation process but also contributes to redefining the possibilities for individuals facing neurological or musculoskeletal impairments.

At the heart of rehabilitation robotics is the ambition to restore mobility and independence for individuals affected by conditions such as stroke, spinal cord injuries, or traumatic brain injuries. Robotic exoskeletons, wearable devices that augment and support body movements, have emerged as instrumental tools in the rehabilitation of lower extremities. By providing powered assistance to leg movements, exoskeletons enable individuals with paralysis or weakness to stand, walk, and engage in weight-bearing activities. These devices not only facilitate physical therapy but also offer psychological benefits, allowing individuals to experience a sense of normalcy and regain a degree of control over their movements.

Upper extremity rehabilitation, crucial for individuals recovering from stroke or traumatic injuries, is addressed through robotic systems designed to enhance arm and hand movements. Robotic exosuits, worn on the arms, assist in repetitive and coordinated movements, promoting muscle reeducation and joint mobility. Hand exoskeletons, equipped with sensors and actuators, facilitate grasp and release functions, allowing individuals with hand impairments to regain essential fine motor skills. The integration of virtual reality and gaming elements further enhances upper limb rehabilitation, turning therapy sessions into engaging and motivating experiences that encourage active participation.

The fusion of robotics with neurorehabilitation principles has given rise to innovative therapies targeting the central nervous system's plasticity. Brain-machine interfaces, a revolutionary development, enable direct communication between the brain and external devices. In the context of rehabilitation, these interfaces open avenues for individuals with severe paralysis to control robotic limbs or external devices using their neural signals. The adaptive nature of the brain allows for neuroplasticity, wherein the brain reorganizes itself to compensate for lost functions, and rehabilitation robotics harnesses this phenomenon to promote recovery and functional adaptation.

Technological advancements in rehabilitation robotics extend beyond the physical realm to address cognitive rehabilitation and neurocognitive disorders. Cognitive rehabilitation robots, often coupled with artificial intelligence, assist individuals with cognitive impairments resulting from conditions like traumatic brain injuries or neurodegenerative disorders. These robots engage users in cognitive exercises, memory tasks, and problem-solving activities, aiming to enhance cognitive functions and support the overall rehabilitation process. The personalized and adaptive nature of these technologies aligns with the individualized needs of users, fostering effective cognitive interventions.

Pediatric rehabilitation, a specialized domain, benefits from the tailored applications of rehabilitation robotics to address the unique challenges faced by children with developmental or neurological disorders. Robotic exoskeletons and wearable devices designed for pediatric use provide support for gait training and movement therapy. Interactive robotic platforms, incorporating play elements and gamified exercises, make rehabilitation sessions engaging and enjoyable for young users. The early integration of rehabilitation robotics in pediatric care not only accelerates physical development but also instills a positive attitude towards therapy in children, setting the foundation for lifelong rehabilitation practices.

The accessibility of rehabilitation robotics is not confined to clinical settings, as home-based robotic systems offer individuals the flexibility to continue rehabilitation beyond traditional therapy sessions. Portable robotic devices, ranging from hand exoskeletons to interactive gaming systems, empower users to engage in consistent and personalized rehabilitation in the comfort of their homes. Tele-rehabilitation platforms leverage technology to connect individuals with remote healthcare providers, allowing for real-time monitoring, guidance, and adjustment of rehabilitation programs. The democratization of rehabilitation through home-based robotics enhances accessibility and continuity of care for individuals with diverse rehabilitation needs.

The integration of artificial intelligence (AI) and machine learning in rehabilitation robotics contributes to personalized and adaptive interventions. AI algorithms analyze user data, including movement patterns, progress metrics, and physiological responses, to tailor rehabilitation programs to individual needs. Machine learning algorithms enable robotic systems to adapt in real-time, optimizing assistance levels, and adjusting to user progress. The synergy of AI and rehabilitation robotics creates dynamic and responsive systems that evolve with users, providing targeted and effective interventions throughout the rehabilitation journey.

Despite the transformative potential of rehabilitation robotics, challenges persist in ensuring widespread adoption and addressing diverse user needs. Cost considerations, often associated with the acquisition and maintenance of robotic systems, pose barriers for some healthcare institutions and individuals. Advocacy for financial support, insurance coverage, and research funding becomes imperative to address these cost challenges and promote equitable access to rehabilitation robotics. Standardization of robotic interfaces and interoperability is another critical consideration to ensure compatibility between different robotic systems and facilitate seamless collaboration among healthcare providers.

User acceptance and engagement are pivotal factors in the success of rehabilitation robotics, emphasizing the importance of user-centered design and collaborative development. Inclusive design principles that consider the diverse needs and preferences of users, including individuals with cognitive or sensory impairments, contribute to the creation of robotic systems that align with the principles of dignity and user autonomy. Continuous feedback loops, involving users in the design and refinement process, foster a sense of ownership and partnership in the use of rehabilitation robotics, enhancing overall user satisfaction and adherence to therapy programs.

The ethical dimensions of rehabilitation robotics revolve around considerations of privacy, consent, and the responsible use of technology in healthcare. As robotic systems collect and process sensitive user data, ensuring robust data security measures and transparent privacy policies becomes imperative. Respecting user autonomy and obtaining informed consent for the use of robotic technologies in rehabilitation are ethical imperatives that underscore the importance of a patient-centered approach. Ethical considerations extend to issues of equity and ensuring that the benefits of rehabilitation robotics are accessible to individuals across diverse socio-economic backgrounds.

In conclusion, rehabilitation robotics stands at the forefront of healthcare innovation, offering a beacon of hope and transformative potential for individuals facing physical and cognitive challenges. From restoring mobility to supporting cognitive functions, these technologies embody the spirit of resilience and human-machine collaboration. As rehabilitation robotics continues to evolve, addressing challenges related to cost, user engagement, and ethical considerations becomes imperative to ensure that the benefits of these technologies are realized by individuals with diverse rehabilitation needs. In the dynamic intersection of technology and rehabilitation, robotics emerges as a powerful ally, empowering individuals to reclaim independence and rewrite the narrative of what is possible in the journey towards recovery and renewed vitality.

CHAPTER VI

Autonomous Systems: Vehicles and Drones

Self-Driving Cars

The advent of self-driving cars represents a paradigm shift in the realm of transportation, promising a future where vehicles navigate the roads with minimal human intervention. Also known as autonomous or driverless cars, these vehicles leverage advanced technologies, including artificial intelligence, sensors, and connectivity, to perceive their surroundings, make decisions, and execute maneuvers. The pursuit of autonomous mobility stems from the vision of enhancing road safety, improving traffic efficiency, and redefining the overall experience of transportation. As self-driving cars transition from futuristic concepts to tangible realities, they bring forth a host of opportunities and challenges that shape the landscape of urban mobility and transportation infrastructure.

One of the primary motivations driving the development of self-driving cars is the aspiration to revolutionize road safety. Autonomous vehicles, equipped with an array of sensors such as lidar, radar, and cameras, possess the ability to perceive their environment with precision and respond to dynamic situations in real-time. The elimination of human errors, a significant contributor to traffic accidents, becomes a focal point in the quest for safer roads. The integration of artificial intelligence enables self-driving cars to analyze complex traffic scenarios, anticipate potential risks, and make split- second decisions to avoid collisions. Proponents argue that a future dominated by autonomous vehicles has the potential to significantly reduce accidents, injuries, and fatalities on the road.

In addition to safety, the promise of increased traffic efficiency and reduced congestion fuels the enthusiasm for self-driving cars. Autonomous vehicles, interconnected through communication technologies, can operate with a level of coordination that surpasses human-driven traffic. Through vehicle-to-vehicle (V2V) communication, self-driving cars can share real-time information about their speed, trajectory, and intended maneuvers. This interconnectedness facilitates smoother traffic flow, minimizes unnecessary stops, and optimizes the use of road infrastructure. Proponents envision a future where self-driving cars navigate through intersections without traffic signals, seamlessly merging into traffic, and collectively contributing to a more efficient and streamlined transportation system.

The advent of self-driving cars is intertwined with the concept of shared mobility and on-demand transportation services. Autonomous vehicle fleets, operated by ride-hailing companies or transportation service providers, promise a transformative shift in how people access transportation. The idea of Mobility as a Service (MaaS) envisions a scenario where individuals can summon a self-driving car through a mobile app, enjoy a comfortable and convenient journey, and disembark without the need for parking. Shared autonomous rides have the potential to reduce the number of vehicles on the road, optimize transportation resources, and contribute to a more sustainable and environmentally friendly urban mobility ecosystem.

The integration of self-driving cars into the transportation landscape prompts considerations about the reshaping of urban infrastructure. Autonomous vehicles, equipped with precise mapping technologies and the ability to communicate with traffic infrastructure, may necessitate adjustments to road designs, traffic signals, and parking infrastructure. Dedicated lanes or zones for autonomous vehicles, coupled with intelligent traffic management systems, could optimize the coexistence of self-driving cars with traditional human-driven vehicles. The potential reduction in the need for parking spaces, as autonomous vehicles can operate continuously and drop off passengers before finding parking remotely, may free up urban space for alternative uses, contributing to more people-centric city planning.

The development and deployment of self-driving cars have garnered attention from major automotive manufacturers, technology companies, and startups alike. Companies such as Tesla, Waymo, Uber, and traditional automakers like General Motors and Ford have invested heavily in autonomous vehicle research and development. The race to perfect self-driving technology has led to significant strides, with some vehicles equipped with advanced driver-assistance systems (ADAS) that offer features like adaptive cruise control, lane-keeping assistance, and automated parking. However, achieving full autonomy, where a vehicle can operate in any scenario without human intervention, remains a complex challenge that involves addressing technological, regulatory, and ethical considerations.

Challenges in the development of self-driving cars extend beyond technical hurdles to encompass regulatory frameworks and public acceptance. The absence of a consistent and standardized regulatory framework poses challenges for the widespread deployment of autonomous vehicles. Different regions and countries have varying laws and regulations regarding autonomous driving, ranging from stringent requirements to more permissive approaches. Harmonizing these regulations to ensure a cohesive and universally accepted set of standards becomes imperative for the seamless integration of self-driving cars into global transportation systems.

Public acceptance and trust are crucial factors influencing the adoption of self-driving cars. While enthusiasts tout the potential benefits of enhanced safety, reduced traffic congestion, and improved accessibility, skeptics and critics express concerns about the ethical implications of autonomous decision-making, data privacy, and the potential job displacement of professional drivers. High-profile incidents involving self-driving car accidents have further fueled skepticism and raised questions about the technology's reliability and safety. Building public trust requires transparent communication about the capabilities and limitations of self-driving cars, rigorous testing and validation processes, and proactive engagement with communities and stakeholders.

Ethical considerations surrounding self-driving cars delve into complex scenarios where decisions must be made in ambiguous situations. The programming of autonomous systems to respond to moral dilemmas, such as choosing between different courses of action in emergency situations, raises profound ethical questions.

Should a self-driving car prioritize the safety of its occupants over pedestrians? How should the decision-making process align with societal values and norms? Addressing these ethical quandaries involves interdisciplinary collaboration, engaging ethicists, policymakers, technologists, and the public in a collective effort to establish ethical guidelines and frameworks that govern the behavior of self-driving cars.

The transition to a future dominated by self-driving cars also has implications for the workforce, particularly those employed in transportation-related roles. The advent of autonomous vehicles may lead to the displacement of professional drivers, affecting occupations such as truck driving and taxi services. The potential societal impact of job displacement requires proactive strategies for retraining and reskilling affected workers, fostering a transition to new industries and occupations. The intersection of automation with workforce dynamics underscores the importance of ethical considerations and responsible deployment of self-driving technology to mitigate negative social and economic consequences.

Cybersecurity emerges as a critical consideration in the era of self-driving cars, where vehicles rely on interconnected systems and communication technologies. The susceptibility of autonomous vehicles to cyber threats, ranging from hacking attempts to malicious interference, necessitates robust cybersecurity measures. Ensuring the integrity of communication networks, securing vehicle-to-infrastructure (V2I) and V2V communication, and implementing safeguards against unauthorized access become paramount to preventing potential cybersecurity vulnerabilities that could compromise the safety and functionality of self-driving cars.

In conclusion, self-driving cars represent a transformative force with the potential to reshape the future of transportation. The promise of enhanced safety, increased efficiency, and improved accessibility underscores the allure of autonomous mobility. However, the journey toward widespread adoption of self-driving cars involves navigating complex challenges, including technological refinement, regulatory alignment, public trust, ethical considerations, and societal impacts. As the automotive and technology industries converge to unlock the full potential of autonomous vehicles, the road ahead involves a careful balance between innovation, responsibility, and collaboration to usher in a future where self-driving cars coexist harmoniously with traditional modes of transportation, offering a new dimension to the way we experience mobility.

Unmanned Aerial Vehicles (UAVs)

Unmanned Aerial Vehicles (UAVs), commonly known as drones, have rapidly evolved from military applications to becoming ubiquitous technological tools with transformative potential across various sectors. These aerial marvels, unburdened by the need for onboard human pilots, have found applications ranging from surveillance and reconnaissance to agriculture, filmmaking, and package delivery. The versatility and accessibility of UAVs have ushered in a new era of innovation, fundamentally changing the way we perceive and interact with the world. As UAV technology continues to advance, the implications for industries, economies, and society at large are far-reaching.

Initially conceived for military purposes, UAVs have transcended their original confines and permeated various civilian domains. In agriculture, UAVs equipped with sensors and cameras offer farmers unprecedented capabilities for crop monitoring, precision agriculture, and yield optimization. These aerial platforms provide real-time data on crop health, soil conditions, and irrigation needs, enabling farmers to make informed decisions that enhance productivity and resource efficiency. The integration of UAVs into agriculture exemplifies their transformative potential in revolutionizing traditional practices and contributing to the sustainability of food production.

In the realm of environmental conservation and research, UAVs serve as invaluable tools for monitoring ecosystems, tracking wildlife, and assessing environmental changes. The ability to access remote or hazardous areas with ease makes UAVs instrumental in gathering data for scientific studies and conservation efforts. From monitoring deforestation and studying marine life to tracking climate patterns, UAVs provide researchers with a cost-effective and efficient means of collecting high-resolution data that contributes to our understanding of the natural world and informs conservation strategies.

The entertainment industry has embraced UAVs for filmmaking and aerial photography, offering filmmakers and content creators unprecedented perspectives and creative possibilities. Equipped with high-quality cameras and stabilization technology, UAVs can capture breathtaking aerial shots, dynamic chase sequences, and panoramic landscapes. The use of drones in filmmaking not only enhances the visual appeal of cinematic productions but also expands storytelling capabilities, allowing directors to achieve shots that were previously logistically challenging or financially prohibitive.

UAVs have revolutionized the way emergency response and disaster management are conducted. In the aftermath of natural disasters or humanitarian crises, drones equipped with thermal imaging cameras, sensors, and communication devices play a vital role in search and rescue operations. These aerial platforms can swiftly navigate disaster-stricken areas, assess the extent of damage, and relay critical information to response teams. The rapid deployment and maneuverability of UAVs enhance the efficiency of emergency response efforts, aiding in saving lives and mitigating the impact of disasters.

The commercial sector has embraced UAVs for logistics and delivery purposes, promising to reshape the landscape of last-mile delivery services. Companies like Amazon and other logistics giants have explored the use of drones to deliver packages quickly and efficiently. The potential advantages include reduced delivery times, lower operational costs, and enhanced accessibility, especially in remote or challenging-to-reach areas. While regulatory challenges and safety concerns remain, the prospect of UAVs revolutionizing the delivery industry underscores their disruptive impact on traditional business models.

The construction and infrastructure sectors have integrated UAVs into surveying, mapping, and inspection processes, offering cost-effective and efficient alternatives to traditional methods. UAVs equipped with LiDAR technology can quickly survey large areas, create detailed topographic maps, and monitor construction sites with unparalleled precision. In infrastructure inspection, drones can access difficult-to-reach structures, such as bridges and towers, to assess their condition, identify potential issues, and streamline maintenance activities. The use of UAVs in these sectors exemplifies how technology can enhance efficiency, reduce costs, and improve overall project management.

While UAVs bring forth a myriad of opportunities, their widespread use has prompted regulatory challenges and concerns related to safety, privacy, and airspace management. Governments worldwide have grappled with developing and refining regulations to ensure the responsible and safe integration of UAVs into airspace. Striking a balance between fostering innovation and safeguarding public safety remains a complex task, as authorities seek to address issues such as air traffic management, pilot certification, and compliance with privacy laws. The development of a regulatory framework that accommodates the evolving landscape of UAV technology is essential to harness its full potential while minimizing risks and ensuring public trust.

Privacy concerns arise from the ability of UAVs to capture images and videos from vantage points that were previously inaccessible. The potential for intrusive surveillance and data collection has spurred discussions about the need for privacy regulations and guidelines governing the use of drones. Addressing these concerns involves a delicate balance between enabling beneficial applications of UAV technology and safeguarding individual privacy rights. Transparent and accountable practices, along with well-defined legal frameworks, are crucial in establishing public trust and ethical use of UAVs across various sectors.

Security considerations have become paramount as UAVs' accessibility has made them potential tools for malicious activities. Unauthorized drone flights near critical infrastructure, public events, or sensitive areas pose security risks that necessitate countermeasures. Governments and security agencies are investing in technologies to detect, track, and mitigate rogue drone activities. The challenge lies in developing effective counter-drone measures without unduly restricting the legitimate and beneficial uses of UAVs. Striking this balance is essential to address security concerns while fostering continued innovation and utilization of UAV technology for societal benefit.

Technological advancements in UAVs continue to push the boundaries of what these aerial platforms can achieve. The development of artificial intelligence and machine learning capabilities enables UAVs to operate autonomously, making decisions based on real-time data and adapting to changing environments. This autonomy enhances their capabilities in tasks such as surveillance, monitoring, and data analysis. The integration of advanced sensors, like hyperspectral and multispectral imaging, enables UAVs to gather more comprehensive and specialized data, opening up new possibilities in fields such as agriculture, environmental science, and infrastructure inspection.

The miniaturization of UAVs, coupled with advancements in battery technology, has led to the creation of nano-drones and swarming capabilities. These miniature UAVs, often capable of flying in coordinated swarms, offer unique advantages in scenarios such as search and rescue missions, environmental monitoring, and surveillance. The ability to deploy a swarm of drones for collaborative tasks extends their utility and effectiveness in scenarios where a single UAV might be limited.

In conclusion, Unmanned Aerial Vehicles have transcended their military origins to become integral tools with transformative potential across diverse sectors. The versatility, accessibility, and evolving capabilities of UAVs continue to reshape industries, offering innovative solutions to age-old challenges. From precision agriculture to emergency response, filmmaking to delivery services, UAVs exemplify the fusion of technology and ingenuity in addressing real-world problems. However, as UAV technology evolves, stakeholders must grapple with regulatory, ethical, and security considerations to ensure responsible and safe integration into society. The ongoing dialogue and collaboration among governments, industries, and the public are vital to harnessing the full potential of UAVs while navigating the complex landscape of challenges and opportunities they present.

Autonomous Marine Systems

Autonomous Marine Systems (AMS) stand at the forefront of technological innovation, heralding a new era in ocean exploration and maritime operations. These unmanned and self-guided marine vehicles, equipped with advanced sensors, artificial intelligence, and communication capabilities, are transforming the way we understand and interact with the vast expanses of the world's oceans. From underwater drones to autonomous surface vessels, AMS are unlocking unprecedented opportunities for scientific research, environmental monitoring, maritime security, and resource exploration.

One of the primary contributions of Autonomous Marine Systems lies in their ability to revolutionize scientific research and exploration of the ocean's depths. Traditional methods of oceanography often involve manned missions, which are limited in duration, costly, and potentially hazardous. AMS, ranging from autonomous underwater vehicles (AUVs) to remotely operated vehicles (ROVs), offer a more efficient and flexible approach. Equipped with a suite of sensors, these vehicles can delve into the ocean's depths, mapping the seafloor, collecting water samples, and capturing high-resolution imagery. The data acquired by AMS contribute to a deeper understanding of oceanography, marine biology, and geology, unraveling the mysteries of the underwater world.

Environmental monitoring and conservation efforts benefit immensely from the deployment of Autonomous Marine Systems. With the capacity to conduct long-duration missions and cover vast areas, AMS enable scientists to monitor ocean health, track changes in temperature, acidity, and salinity, and assess the impact of climate change on marine ecosystems. In addition, AMS are instrumental in studying biodiversity, mapping coral reefs, and monitoring the movements of marine species. The real-time data generated by these autonomous systems contribute crucial insights for marine conservation and the sustainable management of marine resources, facilitating informed decision-making to protect fragile ecosystems.

Maritime security has become an increasingly pressing concern, and AMS play a pivotal role in enhancing surveillance and reconnaissance capabilities. Autonomous surface vessels (ASVs) equipped with advanced sensors and communication systems can patrol maritime borders, monitor shipping lanes, and detect anomalies or potential threats. Underwater drones, capable of covert operations, contribute to the protection of critical infrastructure such as undersea cables and pipelines. The ability of AMS to operate autonomously and cover vast areas of the ocean enhances maritime domain awareness, supporting efforts to combat illegal fishing, piracy, and other illicit activities.

The energy industry benefits from the use of Autonomous Marine Systems in offshore operations, where they offer efficient and cost-effective solutions. AUVs and ROVs are deployed to inspect subsea infrastructure, including oil and gas pipelines, underwater platforms, and renewable energy installations such as offshore wind farms. These autonomous vehicles can navigate complex underwater environments, perform inspections, and relay real-time data to operators on the surface. The deployment of AMS in offshore energy operations enhances safety, reduces operational costs, and minimizes the environmental impact of underwater inspections.

The advancement of technology has given rise to innovative designs and capabilities in Autonomous Marine Systems. Swarming technology, inspired by the collective behavior of marine organisms, enables a group of AMS to work collaboratively on a mission. Swarm robotics, where multiple autonomous vehicles operate in coordination, enhances efficiency and coverage in tasks such as oceanographic surveys or environmental monitoring. The miniaturization of sensors and the use of artificial intelligence enable AMS to adapt to changing conditions, make real-time decisions, and optimize their missions. These technological advancements expand the capabilities of AMS, making them more versatile and effective in addressing a wide range of ocean-related challenges.

Ocean exploration faces numerous challenges, including the vastness of the marine environment, extreme conditions, and the need for sustainable and cost-effective solutions. Autonomous Marine Systems offer a transformative approach to overcoming these challenges, providing a platform for innovative solutions to long-standing problems. Underwater drones equipped with autonomous navigation and obstacle avoidance capabilities navigate challenging underwater terrain, exploring areas that were once inaccessible. ASVs, designed for endurance and adaptability, can operate in a variety of weather conditions, collecting data for extended periods without human intervention. The versatility of AMS positions them as indispensable tools for unlocking the mysteries of the deep sea and addressing urgent global challenges.

Despite the significant advancements, the deployment of Autonomous Marine Systems is not without its challenges. The development of robust communication systems, capable of transmitting data over long distances in challenging underwater conditions, remains a technological hurdle. Power management is another critical consideration, as AMS need to operate for extended periods without requiring frequent recharging or battery replacement. Additionally, ensuring the safety of AMS in shared maritime spaces, preventing collisions with other vessels, and adhering to international regulations pose complex challenges that require careful consideration and collaboration among stakeholders.

The ethical implications of Autonomous Marine Systems also merit attention, particularly in terms of their impact on marine life and ecosystems. While AMS contribute valuable data for scientific research, the potential disturbance to marine animals and habitats raises ethical considerations. Adhering to guidelines that minimize the impact of AMS on marine life, especially in sensitive environments, becomes imperative. Ethical considerations also extend to issues of data privacy, particularly when AMS are deployed for surveillance or monitoring activities near coastal areas or in proximity to human populations.

The future of Autonomous Marine Systems holds exciting possibilities as technology continues to evolve and societal needs drive further innovation. Integrating artificial intelligence with AMS enhances their autonomy and decision-making capabilities, enabling them to adapt to dynamic and unpredictable ocean environments. Further miniaturization of sensors and improvements in energy efficiency will contribute to the development of smaller, more agile AMS that can navigate intricate underwater landscapes and explore areas that were previously inaccessible. The integration of unmanned aerial vehicles (UAVs) with Autonomous Marine Systems creates a comprehensive approach to maritime surveillance and monitoring, allowing for seamless transitions between air and sea.

In conclusion, Autonomous Marine Systems represent a paradigm shift in ocean exploration, offering unprecedented opportunities for scientific research, environmental monitoring, maritime security, and industrial applications. The versatility and adaptability of AMS position them as key enablers for addressing the complex challenges facing the world's oceans. While technological advancements continue to propel the capabilities of AMS, addressing regulatory, ethical, and environmental considerations is crucial to ensuring their responsible and sustainable integration into marine ecosystems. As we navigate the future of ocean exploration, the collaboration between researchers, industry stakeholders, and policymakers becomes essential to harness the full potential of Autonomous Marine Systems for the betterment of our understanding of the oceans and the preservation of these vital ecosystems.

CHAPTER VII

Human-Robot Collaboration

The Rise of Cobots

The landscape of industrial automation is undergoing a profound transformation with the advent of Collaborative Robots, affectionately known as "cobots." Unlike their traditional robotic counterparts, cobots are designed to work alongside human operators, fostering a collaborative and synergistic relationship within the workplace. This paradigm shift in robotics brings forth a new era of human-robot collaboration, where machines enhance human capabilities, improve efficiency, and contribute to safer and more flexible manufacturing processes.

Cobots embody the convergence of cutting-edge robotics and human-centric design, representing a departure from the traditional notion of robots confined to fenced-off areas in factories. The hallmark of cobots is their ability to operate in close proximity to humans, sensing their presence and adapting their behavior accordingly. This proximity enables cobots to engage in intricate tasks, share workspace with human colleagues, and contribute to a more adaptive and responsive manufacturing environment.

One of the primary drivers behind the rise of cobots is their versatility and ease of integration into existing production systems. Traditional industrial robots are often associated with complex programming, specialized environments, and high costs of implementation. In contrast, cobots are designed to be user-friendly, with intuitive programming interfaces that allow non-experts to teach them new tasks. This democratization of robotics empowers small and medium-sized enterprises (SMEs) to leverage automation without the need for extensive expertise, fostering a more inclusive and accessible approach to advanced manufacturing.

Safety takes center stage in the collaborative nature of cobots, addressing concerns that have historically surrounded the integration of robots into shared workspaces. Cobots are equipped with advanced sensors and vision systems that enable them to detect the presence of humans and respond to dynamic changes in their surroundings. These safety features, coupled with the inherent ability to slow down or stop in the presence of humans, minimize the risk of accidents and collisions. As a result, cobots offer a solution to the challenge of ensuring both productivity and worker safety in manufacturing environments.

The agility of cobots is a key factor in their adoption across a diverse range of industries. Unlike traditional robots that often require dedicated fixtures and programming for specific tasks, cobots excel in environments where tasks vary and adaptability is paramount. Cobots can be easily redeployed to perform different tasks, making them ideal for industries with frequent product changes or customization requirements. The ability to quickly reprogram and repurpose cobots aligns with the demands of modern manufacturing, where flexibility and responsiveness to market dynamics are critical for success.
The collaborative nature of cobots extends beyond physical proximity to include a seamless integration of

human and machine decision-making processes. Artificial intelligence and machine learning algorithms empower cobots to learn from human interactions, adapt to changing conditions, and optimize their performance over time. This collaborative intelligence enhances the capabilities of human workers by offloading repetitive and mundane tasks to cobots, allowing human operators to focus on more complex, creative, and value-added aspects of their work.

Cobots find applications across a spectrum of industries, from manufacturing and logistics to healthcare and service sectors. In manufacturing, cobots work alongside human operators in tasks such as assembly, packaging, and quality control. Their ability to handle intricate tasks with precision and repeatability contributes to improved product quality and consistency. In logistics, cobots play a role in material handling, picking, and packing, streamlining warehouse operations and enhancing overall efficiency. The healthcare sector benefits from cobots in tasks such as surgery assistance, patient care, and laboratory operations, where their precision and dexterity complement human skills.

The rise of cobots signals a paradigm shift in the nature of work, prompting discussions about the future of employment and the evolving relationship between humans and machines. Rather than replacing human workers, cobots are envisioned as collaborators, enhancing human capabilities and augmenting the workforce. The introduction of cobots into the workplace necessitates a shift in the skill set required of human workers, emphasizing problem-solving, adaptability, and the ability to collaborate with intelligent machines. This evolution in skills aligns with the changing nature of work in the era of automation, where human-machine collaboration becomes a cornerstone of productivity and innovation.

As cobots become more prevalent in the workforce, the need for standardized safety guidelines and regulations becomes paramount. Ensuring the safe integration of cobots into shared workspaces requires collaboration between industry stakeholders, regulatory bodies, and research institutions. Establishing clear standards for the design, implementation, and operation of cobots contributes to the development of a robust framework that safeguards workers and promotes the responsible adoption of collaborative robotics.

Ethical considerations accompany the rise of cobots, raising questions about the impact on employment, worker well-being, and the potential for unintended consequences. While cobots enhance productivity and contribute to safer work environments, there is a need for thoughtful approaches to address the social and ethical implications. Ensuring fair and equitable access to training opportunities for workers, promoting transparency in the deployment of cobots, and fostering a culture of responsible innovation are essential elements in navigating the ethical dimensions of collaborative robotics.

The future trajectory of cobots is shaped by ongoing technological advancements and the evolving needs of industries. Innovations in artificial intelligence, machine learning, and sensor technologies continue to enhance the capabilities of cobots, enabling them to perform increasingly complex tasks with greater autonomy. The integration of cobots with the Internet of Things (IoT) and connectivity solutions further amplifies their potential, facilitating real-time data exchange and collaboration within smart manufacturing ecosystems.

In conclusion, the rise of cobots marks a transformative moment in the evolution of industrial automation. As collaborative robots become integral members of the workforce, they redefine the dynamics of human-machine collaboration, emphasizing safety, versatility, and adaptability. Cobots embody a vision of the future where automation complements human skills, augments productivity, and contributes to a more inclusive and responsive manufacturing landscape. While challenges and ethical considerations accompany the adoption of cobots, their potential to revolutionize industries and improve the quality of work is undeniable, ushering in an era where humans and machines collaborate harmoniously to shape the future of work.

Ethical Considerations in Human-Robot Interaction

The growing integration of robots into various facets of human life, from workplaces to homes, has ushered in a new era of technological advancements and societal transformations. As robots become increasingly sophisticated and capable of interacting with humans, a host of ethical considerations arises, prompting a careful examination of the impact on individuals, communities, and society at large. The ethical dimensions of human-robot interaction (HRI) encompass a broad spectrum, including issues of privacy, safety, autonomy, accountability, and the potential societal implications of widespread robot adoption.

One of the primary ethical concerns in HRI revolves around privacy, particularly as robots equipped with sensors and cameras become ubiquitous in both public and private spaces. The ability of robots to capture and process visual and auditory information raises questions about the boundaries of surveillance and the protection of personal data. Privacy considerations extend to scenarios where robots are employed in homes for tasks such as caregiving or domestic assistance, necessitating clear guidelines and regulations to safeguard individuals from unwarranted intrusions into their private lives. Striking a balance between the utility of robotic technologies and the preservation of individual privacy becomes a critical challenge in the ethical design and deployment of robotic systems.

Safety is another paramount ethical consideration in HRI, particularly when robots operate in close proximity to humans. The potential for accidents or unintended harm raises questions about the robustness of safety mechanisms and the responsibility of manufacturers and operators. Collaborative robots, designed to work alongside human workers, demand stringent safety standards to prevent collisions, injuries, or other mishaps. Ensuring the reliability of sensors, implementing fail-safe mechanisms, and establishing industry-wide safety guidelines contribute to mitigating the ethical concerns associated with the physical interaction between humans and robots in shared spaces.

Autonomy in robotic systems introduces ethical complexities related to decision-making and accountability. As robots gain the ability to operate autonomously and make decisions based on complex algorithms and artificial intelligence, questions arise about the transparency and accountability of these systems. Understanding the decision-making processes of autonomous robots becomes crucial, especially in scenarios where their actions impact human well-being or have societal implications. Ethical design principles that prioritize explainability and accountability, along with the establishment of legal frameworks, contribute to addressing the challenges of autonomous systems in a responsible and transparent manner.

The potential impact of robots on human autonomy is a nuanced ethical consideration that delves into questions of agency, control, and the potential erosion of individual freedoms. In contexts where robots assist or augment human decision-making, it is essential to ensure that individuals retain the ability to make informed choices and maintain control over critical aspects of their lives. Striking a balance between the empowerment of individuals through robotic assistance and the preservation of human autonomy requires careful consideration of design principles, user consent, and the ethical implications of the decisions made by autonomous systems.

The ethical considerations in HRI extend beyond the immediate interactions between humans and robots to broader societal implications. One of these implications involves the potential displacement of human workers by robotic automation. While robots can contribute to increased efficiency and productivity, concerns arise about the impact on employment and the potential for job displacement. Ethical considerations in this context involve the responsibility of stakeholders, including businesses and policymakers, to address the social and economic consequences of automation. Strategies for reskilling and upskilling the workforce, along with the exploration of alternative employment opportunities, become essential components of ethical decision-making in the adoption of robotic technologies.

Cultural and societal perceptions of robots play a significant role in shaping the ethical landscape of HRI. The design and appearance of robots, along with their roles in various cultural contexts, influence the acceptance and integration of robots into society. Ethical considerations include avoiding the reinforcement of stereotypes or biases through robot design and ensuring cultural sensitivity in the development and deployment of robotic systems. The inclusivity of diverse perspectives in the design and programming of robots contributes to addressing ethical concerns related to cultural representation and the potential impact on societal values.

The ethical use of robots in sensitive contexts, such as healthcare and caregiving, raises unique considerations related to trust, empathy, and the potential for emotional connections between humans and robots. While robots can assist in tasks that enhance the well-being of individuals, the ethical dimensions of emotional engagement and the potential for emotional manipulation require careful consideration. Ensuring transparency about the capabilities and limitations of robots, along with establishing ethical guidelines for their use in emotionally charged contexts, contributes to building trust and maintaining the ethical integrity of human-robot relationships.

As robots become more prevalent in educational settings, ethical considerations emerge regarding their impact on learning outcomes, social interactions, and the potential for reinforcing biases. The use of robots as educational tools raises questions about the equitable distribution of resources, accessibility, and the potential for perpetuating educational inequalities. Ethical guidelines in educational robotics involve ensuring inclusivity, addressing privacy concerns, and promoting educational equity to harness the benefits of robotic technologies for diverse learners.

The development and deployment of autonomous weapons systems, often referred to as lethal autonomous robots, pose profound ethical challenges related to the use of force, accountability, and the potential for unintended consequences. The ethical dimensions of lethal autonomous robots involve considerations about human control over lethal actions, adherence to international humanitarian law, and the prevention of unethical use in armed conflicts. The development of clear ethical frameworks and international agreements becomes imperative to prevent the misuse of robotic technologies in ways that violate human rights and humanitarian principles.

The ongoing evolution of HRI necessitates interdisciplinary collaboration among roboticists, ethicists, policymakers, and the broader public to address the multifaceted ethical considerations. Establishing ethical guidelines and standards for the design, development, and deployment of robots contributes to fostering a responsible and human-centric approach to robotics. Engaging in public discourse and incorporating diverse perspectives ensures that ethical considerations in HRI are reflective of societal values and concerns, promoting a shared understanding of the ethical dimensions of living and working alongside robots.

In conclusion, the rise of robots in various facets of human life brings forth a myriad of ethical considerations that require thoughtful and responsible approaches. From issues of privacy and safety to questions of autonomy, accountability, and societal impact, the ethical dimensions of human-robot interaction demand careful consideration and proactive engagement. By prioritizing transparency, accountability, inclusivity, and ethical design principles, stakeholders can navigate the complex landscape of HRI, fostering a future where robots contribute positively to human well-being while upholding ethical standards that align with our values and aspirations.

Workplace Integration

The workplace of the 21st century is undergoing a profound transformation as technological advancements, automation, and artificial intelligence become integral components of organizational structures. Workplace integration, the seamless collaboration between humans and technology, represents a paradigm shift that brings forth both opportunities and challenges. This section explores the multifaceted dynamics of workplace integration, examining the impact on job roles, organizational culture, productivity, and the evolving nature of work itself.

One of the central aspects of workplace integration is the changing landscape of job roles and skill requirements. Automation and technology are reshaping traditional job functions, automating routine tasks and augmenting human capabilities in complex decision-making and creativity. Jobs that require routine, repetitive tasks are increasingly automated, leading to a redefinition of roles and skillsets. The demand for skills in areas such as digital literacy, data analysis, and problem-solving is rising, reflecting the need for a workforce that can collaborate effectively with technology. Workplace integration necessitates a commitment to continuous learning and upskilling to ensure that employees remain agile and adaptable in the face of technological advancements.

Organizational culture plays a pivotal role in successful workplace integration. The adoption of technology is not merely a matter of incorporating new tools; it requires a cultural shift that embraces innovation, collaboration, and a proactive approach to change. Organizations that foster a culture of openness to technological advancements empower employees to embrace new ways of working and contribute to the integration process. Transparent communication, inclusive decision-making, and a shared vision for the future contribute to a positive organizational culture that supports workplace integration, allowing technology and human capabilities to complement each other seamlessly.

The productivity gains associated with workplace integration are significant, as technology streamlines processes, enhances efficiency, and facilitates collaboration. Automation of routine tasks reduces the likelihood of errors and allows human workers to focus on higher-order thinking, problem-solving, and creativity. Collaborative tools and communication platforms enable real-time information sharing and connectivity, transcending geographical boundaries and time constraints. The synergy between human intuition and the precision of technology creates a dynamic environment where productivity is not merely a result of individual efforts but a collective outcome of human-technology collaboration.

However, workplace integration is not without its challenges. The fear of job displacement and the impact on employment is a prevalent concern as automation becomes more prevalent. It is essential to address these apprehensions through strategic workforce planning, emphasizing the creation of new roles, and providing avenues for reskilling and upskilling. Organizations that proactively manage the human side of workplace integration, fostering a culture of support and empowerment, are better positioned to navigate the challenges associated with technological advancements.

The nature of work itself is evolving in the era of workplace integration. Remote work, flexible schedules, and virtual collaboration have become more prevalent, facilitated by digital technologies. The traditional boundaries of the office space are expanding, giving rise to a more dynamic and distributed workforce. The ability to work from anywhere, coupled with the integration of collaboration tools, not only enhances flexibility but also challenges traditional notions of organizational structures. The rise of the gig economy and the increasing prevalence of freelance work further underscore the transformative impact of workplace integration on the concept of employment.

In addition to job roles and structures, workplace integration also influences employee well-being and satisfaction. The automation of routine tasks can reduce the burden of repetitive work, allowing employees to engage in more meaningful and intellectually stimulating activities. However, the potential for technology to create a constant state of connectivity and blur the boundaries between work and personal life necessitates careful consideration. Organizations must prioritize strategies that promote work-life balance, mental health, and the overall well-being of employees in the integrated workplace to ensure sustained productivity and job satisfaction.

The ethical dimensions of workplace integration come to the forefront as organizations leverage technology to monitor performance, gather data, and make decisions. Balancing the benefits of data-driven insights with the protection of employee privacy requires thoughtful consideration and transparent communication. Establishing ethical guidelines for the use of technology in the workplace, addressing concerns related to surveillance, and ensuring fairness in decision-making processes contribute to a workplace environment that upholds ethical standards in the era of integration.

Collaboration between humans and technology extends beyond routine tasks to include complex problem-solving and decision-making. The field of artificial intelligence (AI) exemplifies this collaborative dynamic, where machine learning algorithms process vast amounts of data to provide insights and augment human decision-making. The ethical considerations in AI-driven decision-making involve issues of transparency, accountability, and bias. Organizations must navigate the ethical complexities by ensuring that AI systems are explainable, auditable, and free from inherent biases that could perpetuate discrimination.

The integration of technology in the workplace also has implications for the physical environment. Smart offices equipped with IoT devices, sensors, and automation systems contribute to energy efficiency, space optimization, and improved resource management. The design of the physical workspace evolves to accommodate collaborative technologies, flexible seating arrangements, and areas that foster creativity and innovation. Workplace integration extends beyond digital tools to include the physical infrastructure that supports a collaborative and technology-driven work environment.

The future trajectory of workplace integration is intertwined with ongoing technological advancements, societal shifts, and the evolving needs of businesses and employees. The integration of augmented reality (AR) and virtual reality (VR) technologies, for example, offers new dimensions to remote collaboration and training. The use of robotics in collaborative tasks and the development of intelligent assistants further exemplify the trajectory toward a more seamlessly integrated workplace. As organizations continue to leverage emerging technologies, it is imperative to stay attuned to the ethical, cultural, and societal implications to ensure a harmonious integration that benefits both humans and technology.

In conclusion, workplace integration is a transformative force that reshapes the dynamics of work, collaboration, and organizational structures. The successful integration of technology requires a strategic approach that addresses the evolving nature of job roles, organizational culture, and the ethical considerations associated with technological advancements. By fostering a culture of adaptability, prioritizing employee well-being, and navigating the ethical dimensions of workplace integration, organizations can create a dynamic and collaborative work environment that leverages the strengths of both humans and technology to achieve shared goals and drive innovation.

CHAPTER VIII

Robotics in Space Exploration

Robotic Rovers and Probes

The exploration of the cosmos has long captured the imagination of humanity, and in recent decades, robotic rovers and probes have emerged as stalwart pioneers, extending our reach into the vast unknown of outer space. These sophisticated machines, equipped with advanced sensors, cameras, and scientific instruments, traverse distant planets, moons, and celestial bodies, unraveling the mysteries of our solar system and beyond. This section delves into the significance of robotic rovers and probes in space exploration, examining their technological evolution, key missions, and the invaluable contributions they make to expanding our understanding of the cosmos.

Robotic rovers stand as marvels of engineering ingenuity, designed to navigate and explore planetary surfaces with precision and autonomy. The epitome of this achievement is NASA's Mars rovers, exemplified by the iconic names Spirit, Opportunity, and Curiosity. These robotic explorers have roamed the Martian landscape, conducting experiments, analyzing rocks and soil, and beaming back breathtaking images of the Red Planet. The technological evolution from the Sojourner rover, which accompanied the Pathfinder mission in the late 1990s, to the advanced capabilities of the Perseverance rover represents a testament to human ingenuity and perseverance in the quest for extraterrestrial knowledge.

The Mars rovers, in particular, have significantly expanded our understanding of the Martian geology, climate, and potential habitability. Instruments on board these rovers have conducted experiments to analyze soil samples, detect organic compounds, and investigate the presence of water. The Curiosity rover, which landed on Mars in 2012, has been a trailblazer in this regard, uncovering evidence of past water flows and offering critical insights into the planet's ancient history. The Perseverance rover, with its sophisticated instruments, continues this legacy, with a specific focus on astrobiology and the search for signs of ancient microbial life.

Probes, on the other hand, extend their exploration beyond planetary surfaces, venturing into the depths of space to study celestial bodies, phenomena, and cosmic structures. The Hubble Space Telescope, launched in 1990, is a paramount example of a space probe that has revolutionized our understanding of the universe. Orbiting Earth, Hubble has captured stunning images of distant galaxies, nebulae, and other astronomical objects, providing astronomers with a wealth of data to unravel the mysteries of cosmic evolution and the nature of dark matter and dark energy.

The Cassini-Huygens mission to Saturn and its moon Titan stands out as another stellar example of the capabilities of space probes. The Cassini spacecraft, equipped with a suite of instruments, orbited Saturn for over 13 years, studying the planet's rings, moons, and magnetic field. The Huygens probe, a part of the mission, descended to the surface of Titan, Saturn's largest moon, providing humanity with its first close-up views of this enigmatic world. The mission not only yielded groundbreaking scientific discoveries, including the identification of liquid methane lakes on Titan, but also captured the public's imagination with awe-inspiring images of the ringed planet and its diverse moons.

The significance of robotic rovers and probes extends beyond the boundaries of our solar system. The Voyager probes, launched in 1977, represent humanity's farthest-reaching emissaries, venturing into interstellar space. These probes, carrying the iconic Golden Records containing sounds and images representing Earth, serve as messengers to potential extraterrestrial civilizations. Voyager 1, in particular, has entered interstellar space, becoming the first human-made object to reach the space between stars. The data transmitted by these probes continue to provide valuable insights into the outer regions of our solar system and the nature of the interstellar medium.

The technological advancements that empower robotic rovers and probes are nothing short of extraordinary. Miniaturization of components, advancements in power systems, and improvements in communication capabilities have all contributed to the enhanced performance and longevity of these robotic explorers. Artificial intelligence and autonomous navigation systems play a crucial role, allowing rovers to make real-time decisions, avoid obstacles, and adapt to unforeseen challenges. These technological strides not only ensure the success of individual missions but also pave the way for future robotic exploration of distant worlds.

Beyond our immediate celestial neighborhood, the search for exoplanets—planets orbiting stars outside our solar system—has become a focal point of space exploration. The Kepler Space Telescope, launched in 2009, and its successor, the Transiting Exoplanet Survey Satellite (TESS), have revolutionized our understanding of the prevalence of exoplanets in the Milky Way galaxy. By employing the transit method, these probes have detected thousands of exoplanets, ranging from Earth- sized rocky planets to gas giants. The data collected by these missions contribute to the identification of potentially habitable exoplanets and advance our quest to answer the age-old question of whether life exists beyond our home planet.

As we marvel at the technological achievements of robotic rovers and probes, their missions also raise ethical considerations. Planetary protection measures are implemented to prevent the contamination of celestial bodies with Earth organisms, reducing the risk of false-positive results in the search for extraterrestrial life. Ethical guidelines also govern the disposal of spacecraft to avoid unintended collisions or interference with future scientific endeavors. The responsible and ethical exploration of space requires a delicate balance between scientific discovery and the preservation of the cosmic environments under investigation.

The synergy between robotic rovers, probes, and human-operated space missions exemplifies the collaborative nature of space exploration. The International Space Station (ISS), a testament to international cooperation, serves as a platform for scientific research, technology development, and human endurance in the space environment. Robotic spacecraft, including the Canadarm2 robotic arm, play crucial roles in the construction, maintenance, and scientific experiments conducted on the ISS. The collaborative endeavors between humans and robotic systems highlight the complementary strengths of both in advancing our understanding of space and conducting experiments that would be impossible on Earth.

Looking ahead, the future of robotic rovers and probes holds exciting possibilities. The upcoming James Webb Space Telescope, set to launch in the near future, promises to revolutionize our observations of the cosmos, peering deeper into space and time than ever before. The Artemis program, aiming to return humans to the Moon, will include robotic rovers as precursors to human missions, conducting scientific experiments and preparing the lunar surface for human exploration. The exploration of icy moons in our solar system, such as Europa and Enceladus, remains a tantalizing prospect, with proposed missions set to deploy robotic probes to explore subsurface oceans that could potentially harbor extraterrestrial life.

In conclusion, robotic rovers and probes stand as vanguards in the grand endeavor of space exploration, pushing the boundaries of our knowledge and expanding the frontiers of human understanding. From the rugged terrains of Mars to the depths of interstellar space, these technological marvels have reshaped our cosmic perspective, unveiling the beauty and complexity of the universe. As we continue to push the limits of technological innovation and scientific discovery, the collaboration between robotic systems and human endeavors remains a cornerstone in our quest to unlock the secrets of the cosmos and answer age-old questions about our place in the universe.

Orbital Robotics

The realm of orbital robotics represents a frontier where cutting-edge technology converges with the vast expanse of outer space, unlocking new possibilities for space exploration and satellite servicing. In Earth's orbit, a dynamic ecosystem of satellites, space probes, and the International Space Station (ISS) orbits, and orbital robotics plays a pivotal role in maintaining, repairing, and augmenting these assets. This section delves into the significance of orbital robotics, exploring its technological foundations, key applications, and the transformative impact it has on our capabilities in the space domain.

At the heart of orbital robotics lies the marriage of advanced robotics and artificial intelligence, enabling machines to perform intricate tasks in the harsh and unforgiving environment of space. These orbital robots, often equipped with robotic arms, manipulators, and sophisticated sensors, can execute precise maneuvers, conduct repairs, and interact with satellites or space station modules with a level of dexterity and accuracy that was once deemed unattainable. The Canadarm2 on the ISS stands as an early exemplar, showcasing the potential of robotic systems in orbit for construction and maintenance tasks.

One of the primary applications of orbital robotics is satellite servicing—a field that has emerged as a game-changer for the sustainability of space assets. Traditionally, when a satellite malfunctioned or reached the end of its operational life, it became space debris, contributing to the growing issue of orbital congestion.

However, with the advent of orbital robotics, satellites can now be serviced and extended in their operational lifespan. Companies like Northrop Grumman's MEV (Mission Extension Vehicle) and Astroscale are pioneering satellite servicing missions, demonstrating the capability to dock with, refuel, and rejuvenate satellites in orbit.

Beyond servicing, orbital robotics plays a crucial role in the deployment and assembly of complex structures in space. The construction of the ISS, a multinational collaborative effort, relied heavily on robotic systems. The aforementioned Canadarm2 and the Special Purpose Dexterous Manipulator (SPDM) were instrumental in assembling and maintaining the various modules of the space station. Looking ahead, orbital robots are envisioned to be key players in the construction of future space habitats, lunar bases, and even interplanetary spacecraft.

The advent of mega-constellations, such as SpaceX's Starlink and OneWeb, which deploy thousands of small satellites for global connectivity, has heightened the importance of orbital robotics for space traffic management. As the number of satellites in orbit increases, the risk of collisions and space debris generation grows. Orbital robots equipped with collision avoidance systems and the ability to deorbit defunct satellites are essential for maintaining a sustainable orbital environment. Companies like Astroscale are developing debris removal missions, where robotic systems will capture and safely dispose of non-functional satellites to mitigate the space debris problem.

The scientific community also reaps the benefits of orbital robotics through servicing and upgrading space telescopes. The Hubble Space Telescope, for instance, underwent several servicing missions where astronauts and robotic systems collaborated to install new instruments, replace aging components, and ensure the longevity of this iconic observatory. As we look toward the deployment of the James Webb Space Telescope, orbital robotics will likely play a role in future servicing missions, ensuring that our observational capabilities in space remain at the forefront of scientific discovery.

The Moon and Mars exploration initiatives further highlight the indispensability of orbital robotics. As humanity contemplates the establishment of lunar bases and prepares for crewed missions to Mars, orbital robots become instrumental in supporting and enabling these ambitious endeavors. From deploying precursor missions to constructing infrastructure in orbit around these celestial bodies, orbital robotics enhances our ability to explore and utilize extraterrestrial environments.

The International Space Station serves as a microcosm of the collaborative potential of orbital robotics. The Canadarm2, operated by astronauts inside the ISS, showcases the synergy between human decision-making and robotic precision. This collaboration extends beyond the station, with robotic arms reaching out to capture resupply spacecraft and perform external maintenance. As we envision future human missions to the Moon and Mars, orbital robotics will continue to play a vital role in supporting human exploration by conducting tasks that are either too dangerous or impractical for human astronauts.

Ethical considerations also permeate the domain of orbital robotics. As we develop capabilities to service, modify, or even dismantle satellites in orbit, questions arise regarding ownership, responsibility, and potential weaponization of these technologies. Establishing international norms and regulations for the responsible use of orbital robotics is paramount to ensure that these capabilities are harnessed for peaceful and collaborative purposes. Ethical guidelines must also address concerns related to space debris mitigation, collision avoidance, and the potential impact of orbital robotics on the long-term sustainability of Earth's orbit.

Looking ahead, orbital robotics is poised to undergo further advancements with the integration of artificial intelligence, machine learning, and advanced sensing technologies. These developments will enhance the autonomy of orbital robots, allowing them to make real-time decisions, adapt to changing conditions, and operate with greater efficiency. The combination of human ingenuity and robotic autonomy creates a powerful synergy that amplifies our capabilities in the exploration and utilization of space.

In conclusion, orbital robotics stands as a transformative force, propelling humanity into a new era of space exploration and satellite servicing. From extending the lifespan of critical space assets to constructing intricate structures in orbit, these robotic systems unlock a realm of possibilities previously relegated to the realm of science fiction. As we venture further into the cosmos, orbital robotics will continue to be a cornerstone of our endeavors, enabling us to reach beyond the confines of Earth and ushering in an era where the vastness of space becomes a realm of collaboration, discovery, and innovation.

Challenges of Space Robotics

As humanity extends its reach into the cosmos through the lens of space exploration, the role of robotics becomes increasingly pivotal. However, this foray into the celestial unknown is not without its formidable challenges. Space robotics, despite its transformative potential, grapples with a host of complexities that demand innovative solutions and careful navigation. This section delves into the multifaceted challenges that confront space robotics, spanning technological hurdles, environmental adversities, ethical considerations, and the intricate dance of human-robot collaboration.

Technological challenges form the bedrock of the hurdles faced by space robotics. The harsh conditions of space, characterized by extreme temperatures, radiation, and vacuum, pose formidable challenges for the design and functionality of robotic systems. Components must endure the rigors of space travel, including the intense vibrations and accelerations during launch. The demand for miniaturization and lightweight design, essential for cost-effective space missions, further complicates the engineering of space robots. The delicate balance between durability and efficiency becomes a central challenge in crafting robotic systems that can withstand the rigors of space exploration.

Communication latency adds another layer of complexity to space robotics. The vast distances between Earth and celestial bodies result in significant communication delays, impacting the ability to control robots in real time. This time lag, often ranging from seconds to minutes, requires robotic systems to possess a degree of autonomy, enabling them to execute tasks and make decisions without constant human intervention. Developing robust autonomous capabilities that align with mission objectives while adapting to unforeseen circumstances remains a persistent challenge in the realm of space robotics.

The dynamic nature of space environments introduces navigation challenges that are distinct from those on Earth. Celestial bodies exhibit irregular terrains, unpredictable gravitational fields, and orbital dynamics that demand adaptive and precise navigation systems. Rovers traversing the surfaces of planets or moons encounter uneven landscapes, obstacles, and varying gravitational forces, necessitating advanced sensing and navigation technologies. Crafting algorithms that enable space robots to navigate autonomously through challenging terrains while avoiding obstacles poses a significant technological hurdle.

Power management emerges as a critical challenge in the realm of space robotics. Traditional power sources, such as solar panels, face limitations in certain environments, such as the shadowed regions of planetary surfaces or during extended missions in deep space. Developing compact, efficient, and resilient power systems that can sustain the energy demands of robotic systems over extended durations becomes imperative. The quest for innovative power solutions, including advanced solar technologies, nuclear power sources, or energy harvesting methods, remains an ongoing challenge for space robotic missions.

As space missions grow in complexity and ambition, the demand for versatile and adaptive robotic systems intensifies. The challenge lies in developing modular and reconfigurable robotic platforms that can fulfill diverse mission objectives. A single robotic system may need to perform tasks ranging from satellite servicing to planetary exploration, demanding a level of adaptability that transcends the capabilities of traditional, specialized robots. The development of modular architectures that enable the customization and reconfiguration of robotic systems for various missions stands as a technological frontier in space robotics.

Beyond technological hurdles, environmental adversities in space present formidable challenges for robotic missions. The corrosive effects of cosmic radiation, micrometeoroid impacts, and the accumulation of space debris pose threats to the longevity and functionality of robotic systems. Crafting materials and protective measures that shield robotic components from these hazards without compromising weight and functionality adds a layer of complexity to space robotics engineering. The imperative to design robotic systems that can endure and operate effectively in the harsh vacuum of space remains a daunting challenge.

The collaboration between humans and robots in space introduces unique challenges related to human-robot interaction and cooperation. The lag in communication, coupled with the complexities of remote operation, demands intuitive and user-friendly interfaces for astronauts. The challenge extends to the development of telepresence technologies that provide astronauts with a sense of immersion and control over robotic systems. Establishing effective communication protocols and interfaces that enable seamless collaboration between humans and robots in space environments is a critical frontier in advancing the capabilities of space robotics. Ethical considerations loom large in the domain of space robotics. As the utilization of robotics in space extends from exploration to resource extraction and potentially even colonization, questions of ownership, environmental impact, and the responsible use of technology come to the forefront. Crafting ethical frameworks that guide the conduct of space missions, prevent the inadvertent contamination of celestial bodies, and ensure the equitable utilization of space resources is an ethical imperative in the evolving landscape of space robotics.

Space debris, the remnants of defunct satellites and spent rocket stages, poses a significant challenge for space robotics. As the density of objects in Earth's orbit increases, the risk of collisions and the generation of more debris grow. Space robots must navigate through this orbital clutter, requiring advanced collision avoidance systems and the ability to adjust trajectories dynamically. The ethical dimension of space debris management adds another layer, demanding responsible disposal methods to minimize the impact on future space missions and orbital sustainability.

International collaboration, while essential for the success of space missions, introduces challenges related to standardization and interoperability in space robotics. Different space agencies and private entities may employ diverse technologies, communication protocols, and interfaces. The challenge lies in establishing common standards that facilitate collaboration, data sharing, and the integration of robotic systems from various sources. Harmonizing international efforts to create a cohesive framework for space robotics cooperation becomes imperative for maximizing the collective potential of global space exploration.

In conclusion, the challenges facing space robotics reflect the complexities inherent in venturing beyond the confines of Earth. Technological, environmental, ethical, and collaborative hurdles define the landscape of space robotics, demanding innovative solutions and interdisciplinary collaboration. As humanity's ambitions in space expand, the evolution of robotic systems to overcome these challenges is both a testament to human ingenuity and a prerequisite for unlocking the full potential of space exploration and utilization. Navigating the intricate dance between technological innovation, environmental resilience, and ethical responsibility paves the way for a future where space robotics becomes an indispensable tool in humanity's quest to explore the cosmos.

CONCLUSION

In the exploration of "Robotics: Engineering the Future," we embarked on a captivating journey through the evolving landscape of robotics—a journey that transcends the realms of technology, engineering, and human ingenuity. This section, with its insightful chapters and comprehensive coverage, has provided a panoramic view of the innovations and developments that define the present and shape the future of robotics.

As we reflect on the diverse facets explored within these digital pages, one resounding theme emerges—the transformative power of robotics. From the foundational principles of robotic engineering to the intricacies of mechanics, dynamics, software, and programming essentials, each chapter unveiled the layers of knowledge that underpin this dynamic field. The exploration of hardware innovations, encompassing advanced sensors, perception systems, actuators, and revolutionary materials, illustrated the continuous pursuit of excellence in creating robots that mirror and augment human capabilities.

The rise of modular robotics showcased the adaptability and versatility that define the next phase of robotic evolution. As we delved into artificial intelligence (AI) integration, machine learning, and cognitive computing, the fusion of robotics with cutting-edge technologies unfolded before us. The synergy between human and machine, illuminated through the chapters on challenges and opportunities, heralded a future where robots become not just tools but collaborative partners in various domains.

The exploration of robotic applications in diverse sectors—from industrial automation and manufacturing to healthcare, logistics, and space—underscored the pervasive impact of robotics on society. Each sector revealed how robotics not only streamlines processes and enhances efficiency but also opens new frontiers of possibility, from surgical precision to autonomous exploration of distant planets.

The ethical considerations woven into the narrative reminded us that as we engineer the future with robotics, a responsible approach is imperative. The discussions on workplace integration, human-robot interaction, and the ethical considerations of autonomous systems underscored the need for a thoughtful and mindful embrace of technological advancements.

In the realm of education, the integration of robotics emerged as a transformative force, shaping young minds, fostering creativity, and preparing the next generation for a future where adaptability and innovation are paramount. The sections on educational robotics programs and the role of robotics in STEM education emphasized the pivotal role of robotics in nurturing skills that extend beyond technical proficiency—skills that are essential for navigating the complexities of a rapidly changing world.

As we draw the curtain on this exploration of robotics, the underlying narrative is one of continuous innovation and limitless potential. The section not only provided a snapshot of the current state of robotics but also illuminated the trajectory ahead. The challenges outlined, whether technological, ethical, or societal, serve as guideposts, beckoning us to navigate the frontier of robotics with a spirit of inquiry, responsibility, and collaborative determination.

In closing, "Robotics: Engineering the Future" is not just a collection of chapters; it is a testament to the relentless pursuit of progress and the boundless possibilities that lie ahead. Whether you are an enthusiast, a student, an educator, or a professional in the field, may this exploration inspire you to delve deeper, question boldly, and contribute to the ongoing narrative of innovation that defines the future of robotics. The journey does not end here—it extends into the uncharted territories of tomorrow, where robotics continues to engineer a future that is as fascinating as it is promising.